Assessment of Technology Education

Editors

Marie Hoepfl
Michael R. Lindstrom

56th Yearbook, 2007
Council on Technology Teacher Education

New York, New York   Columbus, Ohio   Chicago, Illinois   Woodland Hills, California

 **Glencoe**

Copyright © 2007 by the Council on Technology Teacher Education. All rights reserved. Except as permitted under the United States Copyright Act, no part of this publication may be reproduced or distributed in any form or by any means, or stored in a database or retrieval system, without prior written permission of the publisher, Glencoe/McGraw-Hill.

Send all inquiries to:
Glencoe/McGraw-Hill
21600 Oxnard Street
Suite 500
Woodland Hills, CA 91367

ISBN 0-07-877770-4
MHID: 978-0-07-877770-7

Printed in the United States of America.

1 2 3 4 5 6 7 8 9 10   044   11 10 09 08 07

Orders and requests for information about cost and availability of yearbooks should be directed to Glencoe/McGraw-Hill's Order Department, 1-800-334-7344.

Requests to quote portions of yearbooks should be addressed to: Secretary, Council on Technology Teacher Education, in care of Glencoe/McGraw-Hill at the above address, for forwarding to the current secretary.

This publication is available in microform from

UMI
300 North Zeeb Road
Dept. P.R.
Ann Arbor, MI 48106

FOREWORD

This 56th annual yearbook of the Council on Technology Teacher Education continues the yearbook series' uninterrupted tradition of scholarly excellence and promotion of discourse in technology teacher education. In this time of globalization and technological change, we are fortunate to have such a forum in these yearbooks to bring to the fore the kinds of thinking and innovation that have kept our Council strong and vibrant. This year is no exception. Not since 1967 have these yearbooks provided serious and in-depth consideration of the topic of assessment in our field. The tasks of determining what students know by reasoning from evidence; evaluating the effectiveness of educational programs; and coming to understand the cohesion between instruction, curriculum, and assessment represent some of the hardest tasks educators face. Yet we must never shy away from these grand challenges. The Council on Technology Teacher Education wishes to thank and congratulate the co-editors of the 56th yearbook, Dr. Marie Hoepfl and Dr. Michael Lindstrom, for their collective work on assembling this timely yearbook.

Assessments of learning in technology education can provide much-needed information to help educators, policy makers, students, and parents make decisions. This yearbook is well-timed in providing our field with critical insight into aligning assessment in technology education with purpose. The yearbook editors and chapter authors have given careful attention to helping us build an understanding of the diverse approaches to assessment in technology education. The discussion begins with a thorough overview of aligning assessment with purpose and the varied perspectives on assessing technological literacy. Careful consideration of assessment design provides a solid grounding for the reader. From classical test theory to alternative and adaptive assessment designs, the chapter authors provide a good foundation from which to weigh, analyze, and evaluate assessment methodologies. In-depth discussion of large-scale assessment helps the technology educator appreciate the importance and synergy necessary to align curriculum and assessment, while at the same time highlighting the role of accountability in many educational systems. Large-scale assessment is not without peril, and the chapter authors have provided superb insight into developing large-scale assessment of technological literacy.

Assessment of student learning is not an insular enterprise. The chapter authors aid the reader in understanding the nature and scope of assessment in other areas, such as design and technology in the United Kingdom, science, mathematics, and trade and industrial education. The yearbook concludes with a comprehensive chapter on assessment of teachers and programs.

The editors and chapter authors are to be commended for their detailed treatment of the topic of assessment in technology education. We are grateful for their commitment to expanding our perspectives and provoking thought and conversation about the importance of assessment. On behalf of the Council and the Yearbook Committee, we are honored to present this yearbook to the profession. The Council is grateful to have Glencoe/McGraw-Hill as our partner in the yearbook series. Its shared commitment to technology teacher education has made a significant contribution to the field and is truly appreciated. Finally, I join with the Council membership once again in thanking every editor and chapter author who has contributed to this remarkable series of scholarly works since 1952.

Michael A. De Miranda
President, CTTE
March 2007

YEARBOOK PLANNING COMMITTEE

Terms Expiring in 2007
Rodney L. Custer, Chairperson
  Illinois State University
Michael A. De Miranda
  Colorado State University
G. Eugene Martin
  Texas State University, San Marcos

Terms Expiring in 2008
Kurt R. Helgeson
  St. Cloud State University
Linda Rae Markert
  State University of New York at Oswego

Terms Expiring in 2009
Roger B. Hill
  The University of Georgia
Doug Wagner
  Manatee School District, Bradenton, Florida

Terms Expiring in 2010
Mark Sanders
  Virginia Polytechnic Institute and State University
William L. Havice
  Clemson University

Terms Expiring in 2011
Phillip Reed
  Old Dominion University
Brian McAlister
  University of Wisconsin-Stout

OFFICERS OF THE COUNCIL

President
    Michael A. De Miranda
        Colorado State University
        School of Education
        Fort Collins, CO 80523-1584

Vice President
    Marie Hoepfl
        Appalachian State University
        Department of Technology
        Boone, NC 28608

Secretary
    Michael K. Daugherty
        Technology Education
        University of Arkansas
        107 Graduate Education
        Fayetteville, AR 72701

Treasurer
    Brian McAlister
        University of Wisconsin-Stout
        Department of Communication, Education & Training
        Menomonie, WI 54751-0790

Past President
    Rodney L. Custer
        Illinois State University
        Associate Vice President for Research and Graduate Programs
        Normal, IL 61790-4040

YEARBOOK PROPOSALS

Each year at the ITEA International Conference, the CTTE Yearbook Committee reviews the progress of yearbooks to prepare and evaluate proposals for additional yearbooks. Any member is welcome to submit a yearbook proposal, which should be written in sufficient detail for the committee to understand the proposed substance and format. Fifteen copies of the proposal should be sent to the committee chairperson by February 1 of the year in which the conference is to be held. Below are the criteria employed by the committee in making yearbook selections.

CTTE Yearbook Committee

CTTE Yearbook Guidelines

A. **Purpose**
The CTTE Yearbook Series is intended as a vehicle for communicating major topics or issues related to technology teacher education in a structured, formal series that does not duplicate commercial textbook publishing activities.

B. **Yearbook Topic Selection Criteria**
An appropriate yearbook topic should
1. make a direct contribution to the understanding and improvement of technology teacher education.
2. add to the accumulated body of knowledge of technology teacher education and to the field of technology education.
3. not duplicate publishing activities of other professional groups.
4. provide a balanced view of the theme and not promote a single individual's or institution's philosophy or practices.
5. actively seek to upgrade and modernize professional practice in technology teacher education.
6. lend itself to team authorship as opposed to single authorship.

Proper yearbook themes related to technology teacher education may also be structured to
1. discuss and critique points of view that have gained a degree of acceptance by the profession.
2. raise controversial questions in an effort to obtain a national hearing.
3. consider and evaluate a variety of seemingly conflicting trends and statements emanating from several sources.

C. **The Yearbook Proposal**
1. The yearbook proposal should provide adequate detail for the Yearbook Committee to evaluate its merits.
2. The yearbook proposal includes the following elements:
    a) Defines and describes the topic of the yearbook
    b) Identifies the theme and describes the rationale for the theme
    c) Identifies the need for the yearbook and the potential audience or audiences
    d) Explains how the yearbook will advance the technology teacher education profession and technology education in general
    e) Diagrams symbolically the intent of the yearbook
    f) Provides an outline of the yearbook, which includes
        i) a table of contents
        ii) a brief description of the content or purpose of each chapter
        iii) at least a three-level outline for each chapter
        iv) identification of chapter authors(s) and backup authors
        v) an estimated number of pages for each yearbook chapter
        vi) an estimated number of pages for the yearbook (not to exceed 250 pages)
    g) Provides a timeline for completing the yearbook

It is understood that each author of a yearbook chapter will sign a CTTE Editor/Author Agreement and comply with the Agreement. Additional information on yearbook proposals is found on the CTTE Web site at www.ctteonline.org.

PREVIOUSLY PUBLISHED YEARBOOKS

*1. *Inventory Analysis of Industrial Arts Teacher Education Facilities, Personnel and Programs,* 1952.
*2. *Who's Who in Industrial Arts Teacher Education,* 1953.
*3. *Some Components of Current Leadership: Techniques of Selection and Guidance of Graduate Students; An Analysis of Textbook Emphases,* 1954, three studies.
*4. *Superior Practices in Industrial Arts Teacher Education,* 1955.
*5. *Problems and Issues in Industrial Arts Teacher Education,* 1956.
*6. *A Sourcebook of Reading in Education for Use in Industrial Arts and Industrial Arts Teacher Education,* 1957.
*7. *The Accreditation of Industrial Arts Teacher Education,* 1958.
*8. *Planning Industrial Arts Facilities,* 1959. Ralph K. Nair, ed.
*9. *Research in Industrial Arts Education,* 1960. Raymond Van Tassel, ed.
*10. *Graduate Study in Industrial Arts,* 1961. R. P. Norman and R. C. Bohn, eds.
*11. *Essentials of Preservice Preparation,* 1962. Donald G. Lux, ed.
*12. *Action and Thought in Industrial Arts Education,* 1963. E. A. T. Svendsen, ed.
*13. *Classroom Research in Industrial Arts,* 1964. Charles B. Porter, ed.
*14. *Approaches and Procedures in Industrial Arts,* 1965. G. S. Wall, ed.
*15. *Status of Research in Industrial Arts,* 1966. John D. Rowlett, ed.
*16. *Evaluation Guidelines for Contemporary Industrial Arts Programs,* 1967. Lloyd P. Nelson and William T. Sargent, eds.
*17. *A Historical Perspective of Industry,* 1968. Joseph F. Luetkemeyer Jr., ed.
*18. *Industrial Technology Education,* 1969. C. Thomas Dean and N. A. Hauer, eds.; *Who's Who in Industrial Arts Teacher Education,* 1969. John M. Pollock and Charles A. Bunten, eds.
*19. *Industrial Arts for Disadvantaged Youth,* 1970. Ralph O. Gallington, ed.
*20. *Components of Teacher Education,* 1971. W. E. Ray and J. Streichler, eds.
*21. *Industrial Arts for the Early Adolescent,* 1972. Daniel J. Householder, ed.
*22. *Industrial Arts in Senior High Schools,* 1973. Rutherford E. Lockette, ed.
*23. *Industrial Arts for the Elementary School,* 1974. Robert G. Thrower and Robert D. Weber, eds.
*24. *A Guide to the Planning of Industrial Arts Facilities,* 1975. D. E. Moon, ed.
*25. *Future Alternatives for Industrial Arts,* 1976. Lee H. Smalley, ed.
*26. *Competency-Based Industrial Arts Teacher Education,* 1977. Jack C. Brueckman and Stanley E. Brooks, eds.
*27. *Industrial Arts in the Open Access Curriculum,* 1978. L. D. Anderson, ed.
*28. *Industrial Arts Education: Retrospect, Prospect,* 1979. G. Eugene Martin, ed.
*29. *Technology and Society: Interfaces with Industrial Arts,* 1980. Herbert A. Anderson and M. James Benson, eds.
*30. *An Interpretive History of Industrial Arts,* 1981. Richard Barella and Thomas Wright, eds.

*31. *The Contributions of Industrial Arts to Selected Areas of Education,* 1982. Donald Maley and Kendall N. Starkweather, eds.
*32. *The Dynamics of Creative Leadership for Industrial Arts Education,* 1983. Robert E. Wenig and John I. Mathews, eds.
*33. *Affective Learning in Industrial Arts,* 1984. Gerald L. Jennings, ed.
*34. *Perceptual and Psychomotor Learning in Industrial Arts Education,* 1985. John M. Shemick, ed.
*35. *Implementing Technology Education,* 1986. Ronald E. Jones and John R. Wright, eds.
 36. *Conducting Technical Research,* 1987. Everett N. Israel and R. Thomas Wright, eds.
*37. *Instructional Strategies for Technology Education,* 1988. William H. Kemp and Anthony E. Schwaller, eds.
*38. *Technology Student Organizations,* 1989. M. Roger Betts and Arvid W. Van Dyke, eds.
*39. *Communication in Technology Education,* 1990. Jane A. Liedtke, ed.
*40. *Technological Literacy,* 1991. Michael J. Dyrenfurth and Michael R. Kozak, eds.
 41. *Transportation in Technology Education,* 1992. John R. Wright and Stanley Komacek, eds.
*42. *Manufacturing in Technology Education,* 1993. Richard D. Seymour and Ray L. Shackelford, eds.
*43. *Construction in Technology Education,* 1994. Jack W. Wescott and Richard M. Henak, eds.
 44. *Foundations of Technology Education,* 1995. G. Eugene Martin, ed.
*45. *Technology and the Quality of Life,* 1996. Rodney L. Custer and A. Emerson Wiens, eds.
 46. *Elementary School Technology Education,* 1997. James J. Kirkwood and Patrick N. Foster, eds.
 47. *Diversity in Technology Education,* 1998. Betty L. Rider, ed.
 48. *Advancing Professionalism in Technology Education,* 1999. Anthony F. Gilberti and David L. Rouch, eds.
*49. *Technology Education for the 21st Century: A Collection of Essays,* 2000. G. Eugene Martin, ed.
 50. *Appropriate Technology for Sustainable Living,* 2001. Robert C. Wicklein, ed.
 51. *Standards for Technological Literacy: The Role of Teacher Education,* 2002. John M. Ritz, William E. Dugger, and Everett N. Israel, eds.
 52. *Selecting Instructional Strategies for Technology Education,* 2003. Kurt R. Helgeson and Anthony E. Schwaller, eds.
 53. *Ethics for Citizenship in a Technological World,* 2004. Roger B. Hill, ed.
 54. *Distance and Distributed Learning Environments: Perspectives and Strategies,* 2005. William L. Havice and Pamela A. Havice, eds.
 55. *International Technology Teacher Education,* 2006. P. John Williams, ed.

* Out-of-print yearbooks can be obtained on microfilm and in printed copies. For information on price and delivery, write to UMI, 300 North Zeeb Road, Dept. P.R., Ann Arbor, Michigan 48106.

PREFACE

Assessments provide the data from which evaluations of student progress are derived. They produce the evidence demonstrating whether educational goals have been met. Assessment is a critical attribute of educational accountability and defines what is ultimately valued within a course or program. We believe, therefore, that it is only appropriate for a CTTE yearbook to purposefully address assessment within the context of technology education.

We have attempted to identify and include the essence of quality assessment practices commonly found in educational domains and view them through the lens of a technology educator. Expertise in these areas was sought from within the ranks of technology educators where possible, and every attempt was made to frame examples from the perspective of technology education programs and classrooms. We hope the result is a text that holds true to best educational assessment practices and at the same time is customized for direct application within technology education programs.

We have also tried to stay true to the ITEA vision of assessment found within the *Standards for Technological Literacy: Content for the Study of Technology* and the compendium document *Advancing Excellence in Technological Literacy: Student Assessment, Professional Development and Program Standards.*

Finally, we are grateful to have had the opportunity to contribute to a profession in which we so strongly believe. It has been an extraordinary opportunity to review and assemble the thoughts from a number of respected authors. In spite of the hard work involved, it has been a pleasure to be associated with Glencoe/McGraw-Hill, the Yearbook Committee, the authors, and the process of this yearbook.

56th Yearbook Editors
Marie Hoepfl
Michael R. Lindstrom

ACKNOWLEDGMENTS

We would like to thank our friends at Glencoe/McGraw-Hill for their ongoing commitment to technology education through the CTTE yearbook series. The yearbooks serve to document the key issues in our profession today and provide a valuable resource in establishing and communicating a future direction.

We would also like to thank the Yearbook Committee members for their guidance and support during the planning and writing processes. Their willingness to share expertise as well as their commitment to a quality product has resulted in a series of yearbooks that has significantly benefited the profession. With the Committee's assistance, we hope that we have extended that tradition.

We also need to recognize the contribution and essential attention to detail provided by our proofreaders, Karen Ginn and Kay Dickson from Appalachian State University.

Finally, we hold the author team in highest regard. We thank the authors for meeting aggressive timelines and demonstrating flexibility. Most importantly, we thank them for sharing their insights and expertise on the critical topic of technology education assessment. We believe that the profession has long needed to address this topic.

Dr. Jared V. Berrett is an assistant professor in the Technology Teacher Education (TTE) program at Brigham Young University. He holds bachelor's degrees in psychology and technology education, a master's degree in technology education, and a PhD in educational psychology from the University of Illinois. He currently serves as the graduate coordinator for the TTE program, is president of the Technology Education for Children Council, and enjoys working with his colleagues at the National Center for Engineering and Technology Education. Berrett has taught technology in four high schools and keeps an active presence teaching in grades K–8 for his research. He is passionate about teaching and learning, and is dedicated to understanding how to help today's youth become technologically literate while developing creative abilities and higher-order cognitive thinking skills.

Dr. Janet Z. Burns received a BS in home economics journalism from the University of Wisconsin-Madison, and earned her MS and PhD in human resource development with a cognate in vocational and career development at Georgia State University. She has been involved as a faculty member in trade and industrial education for 10 years and is currently an associate professor in the Department of Middle/Secondary Education and Instructional Technology at Georgia State University. She is the unit coordinator for the Technology/Career Education program.

Burns was a trainer for United Airlines, an instructional designer for Equifax Services, and has served as an education consultant to various school districts in the metro Atlanta area. She has also served as the research chair for the National Association of Industrial and Technical Teacher Educators (NAITTE) and is editor of the *Journal of Industrial Teacher Education*, Vols. 42–43. She has memberships in ACTE, ACTER, the Academy of Human Resource Development, and the American Psychological Association.

Dr. Phillip Cardon received a BS in automotive technology from Weber State University, an MS in technology education from Brigham Young University, and a PhD in technology education from The Ohio State University. He has taught technology education at the postsecondary level since 1999 and is currently an associate professor at Eastern Michigan University. Cardon has completed research in the area of curriculum development and special-needs populations in technology education. His teaching interests include methods and curriculum development and technology for children. His service to the technology education field includes work in the CTTE, Ohio Technology Education Association, ITEA, Learning Institute for Technology Education (LITE), and service as an EPT co-trustee.

Dr. Rodney L. Custer is associate vice president for Research, Graduate Studies, and International Education at Illinois State University. He is currently serving as past president of the Council on Technology Teacher Education and has been actively involved in a number of technology education initiatives, including the development of the *Standards for Technological Literacy*. He also currently serves as chair of the CTTE Yearbook Committee. Custer has been active with committee work for the National Research Council, working on topics related to the advancement of technological literacy, including *Technically Speaking* and *Tech Tally*. He has served as a program officer for the National Science Foundation in Washington, DC, and currently serves as a consultant and external evaluator for several NSF-funded projects. His research interests are concentrated in technological problem solving and integrated learning.

Dr. Gerald Day is the graduate studies coordinator for the Department of Technology, University of Maryland Eastern Shore. Previously, Day was with the Maryland State Department of Education (MSDE) for 23 years in a variety of positions. His last position at MSDE was as accountability director for the Division of Career Technology and Adult Learning. Day has been involved in designing program evaluation systems at the state and local levels throughout his career. He has taught at the middle school, high school, and college levels for 35 years. Day is currently the NCATE/ITEA/CTTE program review coordinator.

Dr. Ruth Xiaoqing Guo completed her doctoral degree at the University of British Columbia in 2006. Her dissertation research involved a large-case assessment of literacy in teacher education, titled "Information and Communications Technology (ICT) Literacy in Teacher Education: A Case Study of the University of British Columbia." She has published works on literacy and gender in technology, and is interested in how ICT literacy is sustained by teachers during their first five years of teaching. She began teaching at the University of Ottawa in September 2006.

Dr. Jim Haynie received his BS in industrial arts education from Old Dominion University, an MS in industrial education from Clemson University, and a PhD in curriculum and instruction with a minor in educational psychology from Pennsylvania State University. He has taught industrial arts, technology education, educational psychology, and tests and measurements. His teaching experience includes six years in middle and high schools—and 26 years at the university level. Haynie is currently professor and coordinator of the technology education program at North Carolina State University. He has authored 12 refereed journal articles on classroom testing and assessment as well as monographs on student evaluation and program assessment. He is widely consulted on test development, assessment of evaluation instruments, and evaluation issues.

Dr. Kurt R. Helgeson is currently associate professor and chair of the Department of Environmental and Technological Studies at St. Cloud State University, where he has taught since 1998. He teaches a variety of classes in the areas of construction and technology education, as well as several graduate courses. Helgeson received his BS and MS degrees from St. Cloud State University and an EdD in technology education from West Virginia

University. He is a member of CTTE, MTEA, ITEA, and the Minnesota Firefighter Instructors Association. Helgeson was co-editor of the 2003 CTTE Yearbook, *Selecting Instructional Strategies for Technology Education.* Prior to working at St. Cloud State University he worked in the kitchen-cabinet industry for five years as a design and quality assurance manager.

Dr. Marie Hoepfl received a BS in industrial education at Miami University of Ohio, an MS in education administration, and an EdD in technology education at West Virginia University. She taught at the middle and high school levels for six years and has taught at the university level since 1994. Currently, she is professor and graduate program director in the Department of Technology at Appalachian State University. Hoepfl was a program officer at the National Science Foundation in 2001–2002, served two terms as treasurer of the CTTE, and is currently vice president of the CTTE. She has published several articles on the topic of assessment and evaluation, and recently served as external evaluator for a pilot study on the use of performance assessment for accountability purposes for the state of North Carolina.

Mr. Richard Kimbell was the first professor of technology education at London University. He has taught technology in schools and has been course director for undergraduate and postgraduate courses of teacher education. He launched the doctoral program in technology education at Goldsmiths College and has supervised many graduates of that program. From 1985 to 1991, Kimbell directed the Assessment of Performance Unit (APU) project for the Department of Education and Science, which was charged with assessing the technological capability of 10,000 pupils in 700 schools throughout England, Wales, and Northern Ireland. In 1990, he founded the Technology Education Research Unit (TERU) as the base from which to manage his expanding research portfolio. In the subsequent period, research sponsors have included research councils (e.g., ESRC, NSF [USA]), industry (e.g., LEGO, BP), government departments (e.g., DfES, DfID), as well as professional and charitable organizations (e.g., the Engineering Council, Royal Society of Arts, and the Design Museum).

This research has been analyzed in Kimbell's book, *Researching Design Learning,* co-authored with Kay Stables (Springer: December 2006). His previous book, *Assessing Technology,* won the Outstanding Publication Award (1999) from the Council on Technology Teacher Education (CTTE) at the International Technology Education Association Conference in Minneapolis, MN. He has been commissioned to write reports for the British Council, the U.S. Congress, UNESCO, and NATO. He is currently a consultant to the National Academy of Engineering in Washington, offering recommendations for ways to approach technological literacy assessment. He has written and presented television programs, and he regularly lectures internationally.

Dr. Michael R. Lindstrom received BS degrees in mathematics and industrial education from Winona State University, an MS in industrial education from Winona State University, and an EdD in work, community, and family education from the University of Minnesota. He has taught industrial technology and technology education at the middle, high school, and university levels for 21 years. Lindstrom was a technology education consultant for the Anoka-Hennepin Schools (MN) from 1991 to 1993, and from 1993 to 1999 served the district as an instructional facilitator. Currently, Lindstrom is assessment facilitator for the Anoka-Hennepin Schools. He is a past president and current secretary for the North Suburban Technology Education Association. Lindstrom is past president and current professional growth chair and conference program chair for the Minnesota Technology Education Association. He served as the executive secretary of the Council of Minnesota Professional Education Associations, and is currently the past chair of the SciMathMN organization. Lindstrom holds memberships in NSTEA, MTEA, ITEA, MCTM, CTTE, Epsilon Pi Tau, and Phi Kappa Phi.

Mr. Greg Pearson is a program officer with the National Academy of Engineering (NAE), where he directs the academy's efforts related to technological literacy. In this capacity, Pearson most recently served as the responsible staff officer for the Committee on Assessing Technological Literacy, a joint project of the NAE and the National Research Council. The committee's final report, Tech Tally: Approaches to Assessing Technological Literacy, was published in summer 2006. He was co-editor of Technically Speaking: Why All Americans Need to Know More About Technology (2002), the result of another Academies project, and he oversaw a review of national K–12 content standards for the study of technology developed by the International Technology Education Association. He has worked collaboratively with colleagues within and outside the National Academies on a variety of projects involving K–12 science, mathematics, technology, and engineering education, and the public understanding of engineering and science.

Dr. Stephen Petrina is an associate professor in the Department of Curriculum Studies at the University of British Columbia, where he coordinates the technology studies program. Petrina has published widely in the area of design and technology education. He recently published a major synthesis of the history of educational technology and media in the spring 2002 issue of History of Education Quarterly and is publishing a collection of essays titled The Critical Theory of Design and Technology Education. His new textbook, Advanced Teaching Methods for the Technology Classroom, was published in September 2006. Petrina is also directing a major study on cognition and technology ("How We Learn: Technology Across the Lifespan").

Dr. Maria Araceli Ruiz-Primo holds a PhD in educational psychology from the University of California, Santa Barbara. She was co-director of the Stanford Education Assessment Laboratory until she moved to the University of Colorado at Boulder as a senior researcher in the School of Education's Research and Evaluation Methodology program. She is currently director of the Research Center and associate professor in the School of Education, University of Colorado at Denver. Ruiz-Primo's research focuses on educational measurement and the development and technical evaluation of innovative science learning assessment tools, including performance tasks, concept maps, and students science notebooks. She has conducted research on the instructional sensitivity of assessments and their proximity to the enacted curriculum. She participated in the development of the science teacher certification assessment of National Board of Professional Teaching Standards. Ruiz-Primo was awarded a fellowship from the American Educational Research Association in 1993, and from 2001 to 2002 was a program officer in the National Science Foundation's Research, Evaluation, and Communication (REC) and Elementary, Secondary, and Informal Education (ESIE) divisions. She is the first author of the student guide *Statistical Reasoning for the Behavioral Sciences*.

Ms. Karen M. Schaefer holds an MA in mathematics from Georgia State University. She has been a secondary school teacher and has taught in both the Department of Mathematics and the Department of Middle/Secondary Education and Instructional Technology at Georgia State University. Before her retirement, she taught and supervised new teachers in the Technology/Career Education program.

Dr. Anthony E. Schwaller has been working in the field of technology education for 35 years. During his tenure in technology education, he received the Distinguished Technology Educator Award in 1990, the Special Recognition Award from ITEA in 1995, and the Academy of Fellows Award from the ITEA in 2005. Schwaller was selected as Technology Teacher Educator of the Year in 1997 by the Council on Technology Teacher Education, and he received the Outstanding Research Award in 2003 for his research on modular technology and the *Standards for Technological Literacy*. Schwaller has written more than 50 articles in professional journals; has written 10 textbooks in the field of technology, instructional strategies, and technology education; and has given more than 50 presentations on topics such as energy, transportation, instructional strategies, leadership, and accreditation/assessment. In addition, he has helped to facilitate the process of revising the ITEA/CTTE/NCATE curriculum standards for technology education. Schwaller taught at St. Cloud State University for 28 years and was department chair for 15 of those years. He retired in May 2006.

Mr. Tom Shown is a technology education supervisor for the North Carolina Department of Public Instruction. He has served in this position for more than 15 years and is a former technology education teacher. He has a BA in philosophy from the University of Southern Colorado and an MA in technology education from North Carolina State University.

Dr. Steven L. Shumway is an assistant professor in the Technology Teacher Education Program in the School of Technology at Brigham Young University. Prior to completing a doctoral degree in 1999 at Utah State University, Shumway taught high school electronics/technology education classes for six years. He maintains a close relationship with public high schools and enjoys opportunities to interact with high school students and teachers. At Brigham Young University, Dr. Shumway's primary responsibilities include teaching graduate and undergraduate classes, supervising student teaching, serving as the advisor for BYU's chapter of the Technology Education Collegiate Association, and serving as the Technology Teacher Education Program chair. Nationally, Shumway was a standards team member for Advancing Excellence in Technological Literacy (2003) and is currently an ITEA standards specialist assisting technology educators in advancing technological literacy standards-based reform.

Ms. Rhonda Welfare is a consultant for planning and performance management for the North Carolina Department of Public Instruction, where she works primarily with analysis of performance results for local, state, and federal career and technical education accountability. She received a BS in journalism from Northwestern University and an MS in training and development from North Carolina State University, where she is currently pursuing a doctoral degree in adult and higher education.

TABLE OF CONTENTS

Foreword iii
Yearbook Planning Committee iv
Officers of the Council v
Yearbook Proposals vi
Previously Published Yearbooks vii
Preface ix
Acknowledgments x

SECTION 1. OVERVIEW OF ASSESSMENT IN TECHNOLOGY EDUCATION
**Chapter 1: Aligning Assessment with Purpose: When, What,
    and How to Assess** 1
    Marie Hoepfl
    Appalachian State University
    Michael R. Lindstrom
    Anoka-Hennepin Schools

**Chapter 2: Assessing Technological Literacy: A National
    Academies Perspective** 17
    Rodney L. Custer
    Illinois State University
    Greg Pearson
    National Academy of Engineering

SECTION 2. ASSESSMENT DESIGN
**Chapter 3: Conventional Classroom Assessment Tools: Item Selection, Development, and
    Evaluation** 33
    W. J. Haynie III
    North Carolina State University

**Chapter 4: Alternative Classroom Assessment Tools and
    Scoring Mechanisms** 65
    Marie Hoepfl
    Appalachian State University

Chapter 5: Assessing Students with Disabilities in Technology Education 87
    Phillip L. Cardon
    Eastern Michigan University

Chapter 6: Data Analysis Techniques 103
    Michael R. Lindstrom
    Anoka-Hennepin Schools

Chapter 7: Applying Classroom Assessment Data: Communicating Results
 and Modifying Instruction..125
 Jared L. Berrett
 Steven Shumway
 Brigham Young University

SECTION 3. LARGE-SCALE ASSESSMENTS
Chapter 8: Aligning Curriculum and Assessment in the Development of a Statewide
 Accountability System 139
 Rhonda Welfare
 Thomas Shown
 North Carolina Department of Public Instruction

Chapter 9: Developing a Large-Scale Assessment of
 Technological Literacy..157
 Stephen Petrina
 University of British Columbia
 Ruth Xiaoqing Guo
 University of Ottawa

Chapter 10: Assessment of Design and Technology in the U.K.: International Approaches to
 Assessment 181
 Richard Kimbell
 Technology Education Research Unit (TERU)
 Goldsmiths College, University of London

SECTION 4. ASSESSMENT IN OTHER FIELDS OF STUDY
Chapter 11: Assessment in Science and Mathematics: Lessons Learned........................203
 Maria Araceli Ruiz-Primo
 University of Colorado at Denver/Stanford Education Assessment Laboratory

Chapter 12: Skills Assessment in Trade and Industrial Education 233
 Janet Z. Burns
 Karen M. Schaefer
 Georgia State University

SECTION 5. ASSESSMENT OF TEACHERS AND PROGRAMS
Chapter 13: Conducting Program Assessments..251
 Gerald F. Day
 University of Maryland Eastern Shore
 Anthony E. Schwaller
 St. Cloud State University

Chapter 14: Assessing Teacher Readiness and Teacher Performance 273
 Kurt R. Helgeson
 St. Cloud State University

Aligning Assessment with Purpose: When, What, and How to Assess

Chapter 1

Marie Hoepfl
Appalachian State University

Michael R. Lindstrom
Anoka-Hennepin Schools

INTRODUCTION

Assessment is a fundamental part of the education process, yet its usefulness has been undermined by two influences. First, many educators view educational measurement as a specialized activity that is difficult to understand and, therefore, as something that should be carried out by assessment experts. Second, the appropriation of assessment as the key component of the so-called accountability movement in schools has, in many cases, meant that assessments are "educationally destructive" rather than "instructionally illuminating" (Popham, 2001, p. 27).

The editors of the 56th Council on Technology Teacher Education yearbook have attempted to create a guide to producing instructionally illuminating assessments for technology education. Although a plethora of assessment resources exists for educators, few are specifically focused on the unique aspects of the technology education classroom. This book is intended to fill that gap, by customizing the topic of assessment for technology education instructors and programs, at both the K–12 and post-secondary levels. Administrators of technology education programs and preservice technology teachers will also find topics in many of the chapters that are appropriate for their needs.

PURPOSE OF THE YEARBOOK

The purpose of the 2007 CTTE yearbook is to select and summarize the aspects of assessment that are most critical and most applicable to technology education programs. While it cannot be considered the "unabridged" version of technology education assessment, we hope that this yearbook captures the key issues and provides useful references for those interested in further study on any of these assessment topics.

Any book focused on a common theme that incorporates the work of more than a dozen authors will present editorial challenges. These challenges include achieving a common style, linking the chapter topics in a meaningful way, and eliminating redundancy. Although we sought to address these challenges, we also realized that end users of CTTE yearbooks frequently use single chapters in isolation from the remainder of the text. With this in mind, we eliminated redundancy where possible, while at the same time maintaining sufficient overlap of topics to allow chapters to stand alone.

OVERVIEW OF CONTRIBUTORS AND CHAPTERS

As this yearbook was being envisioned, we worked to identify the assessment topics with greatest potential for positively impacting technology education teachers and programs. Next, authors with expertise in each topic were sought, and the topics were assembled into individual chapters and sections.

Section 1 provides an overview of assessment in technology education. We attempt to outline the major themes within technology education assessment, and then discuss several issues that must be considered to maximize the benefits of assessment. Custer and Pearson provide a more broad-based look at assessing technological literacy, based on their work at the National Academies that has resulted in the recent publication of the book *Tech Tally* (National Academy of Engineering and National Research Council, 2006).

Section 2 focuses on the design of assessments and includes several how-to chapters detailing design considerations for a variety of assessment tools. For example, Haynie describes strategies for creating traditional classroom assessment tools (tests and quizzes) and provides a number of examples to illustrate how these strategies are applied in the technology education classroom. Hoepfl gives a similar treatment to the design of alternative assessment tools. Cardon examines some of the issues surrounding assessment of students with disabilities in technology education. Lindstrom looks at assessment more from the district perspective by describing methods of dealing with data sets and for drawing conclusions from data. Finally, Berrett and Shumway describe ways of communicating and using the results of assessments.

Section 3 looks at large-scale assessments and provides case studies of two educational settings where large-scale assessments of technology education have been adopted. Welfare and Shown describe the assessment system adopted in North Carolina, including how the assessments are structured, how the data are used, and how the assessment system has evolved since its inception. Kimbell details the process of developing a large-scale performance assessment of design and technology in the United Kingdom. Petrina and Guo discuss the challenges inherent in developing large-scale assessments and essentially issue a challenge to the technology education community.

In Section 4, the authors focus our attention on assessment in other fields of study. Ruiz-Primo describes her and her colleagues' work on assessment in science and mathematics, and shares mechanisms for systematically thinking about, and measuring, different types of knowledge within any discipline. Burns and Schaeffer examine assessment issues within the broad field of trade and industry education. They frame their discussion around the need within trade and industry education to align assessment with identified industry skills and demands and, in some cases, with professional certification requirements.

Section 5 examines issues related to the assessment of teachers and programs. Day and Schwaller identify several organizations that assess teacher education programs, including the National Council for Accreditation of Teacher Education (NCATE). They describe data-collection strategies for program assessment, as well as strategies for applying assessment data toward the goal of program improvement. Finally, Helgeson discusses the assessment of teaching performance at both the preservice and in-service levels. He uses a case-study approach to illustrate how one institution has documented the performance of its preservice teachers.

Critical Attributes for Chapters

Based, in part, on guidance provided by the CTTE Yearbook Committee, we developed a list of critical attributes for all yearbook chapters. Authors were asked to consider how these attributes could be incorporated into their chapters.

First and foremost, we asked that chapter material focus on the technology education context and ways that assessment is uniquely applied within that context. Wherever possible, examples used to illustrate ideas were expected to come from the technology education classroom.

Second, we wanted chapter material to emphasize strategies for using assessments to inform instruction. This includes designing assessment tools to ensure the collection of good data, as well as applying the information collected. This criterion also related to a third attribute: communication of results. Communicating results includes identifying the audience for the assessment (students, parents, teachers, administrators, communities, legislators); reporting or communicating the results to that audience; and doing so with clarity, accuracy, and effective timing.

We asked that each chapter, as appropriate, address the politics of assessment—in other words, the effect that political trends have on assessment and the influence of assessment on educational policy. However, it was suggested that there should not be a preponderance of references to specific pieces of legislation, such as the No Child Left Behind Act, because of their temporal and place-specific nature. Rather, such legislation was to be discussed in relation to general educational trends or to its lasting legacies (e.g., the United States' Education for All Handicapped Children's Act of 1975).

We asked that selected chapters include material related to trends such as performance-based assessment and computer-based assessment. Both of these assessment "innovations" have particular applicability in the technology education context. Finally, we wanted chapter authors to make obvious the fact that technological literacy is the goal of technology education programs. One of the ways this was achieved was by including references to the *Standards for Technological Literacy* (STL), *Advancing Excellence in Technological Literacy* (AETL), *Measuring Progress*, and other ITEA publications, where appropriate (International Technology Education Association, 2000, 2003, 2004).

MAJOR ISSUES IN ASSESSMENT

The age of accountability in education has brought assessment to the forefront of almost every aspect of teaching and learning. This emphasis on accountability, however, has resulted in what Stiggins (2004) has called a "naïve and counterproductive assessment legacy" (p. 23). Ideally, educational programs will exhibit a close linkage between the three central elements of the educational process—curriculum, instruction, and assessment—but in reality there is often little coherence between the components of this triad (Pellegrino, 2002). This lack of alignment may stem, in part, from outdated perceptions about how people learn, as well as from the misguided application of accountability testing, which can create an unbalanced enterprise (Pellegrino, Chudowsky, and Glaser, 2001; Kohn, 2000). It is ironic that, as states are hailed for developing standardized tests to better enable the tracking and comparison of educational programs, the use of classroom-based assessments has been de-emphasized. In other words, classroom teachers have been increasingly cut out of the assessment loop, thus making it less likely that they will use formative assessment for instructional improvement, a process that has been convincingly demonstrated to be effective for improving learning (Black et al., 2004).

Technology education is not immune to these trends, and in addition, it encounters further challenges. Technology education courses are often performance-based in nature, which makes the collection of assessment data in these courses unique from many other content areas. Few nationally recognized standardized assessments focus on the topic of technology education, and courses are seldom textbook-based, which fosters a wider range of topics and foci than would be found in courses such as algebra and biology. In spite of frequent references within the literature to the need for better "STEM" education—science, technology, engineering, and mathematics—an understanding of the role that technology plays in this community of disciplines has not been demonstrated. The lack of national emphasis on this arguably critical area of learning has meant that few resources have been devoted to defining that role.

It may seem a remote possibility that technology education will ever fall under the umbrella of high-stakes testing. However, based on the ever-increasing societal dependence on technology and technological literacy, a case could easily be made that the United States and other nations in fact need high-stakes testing of technological literacy in all schools. Although development of large-scale, potentially high-stakes technology education assessments is not guaranteed to propel technology education into the mainstream of core educational topics, large-scale assessments do remain one of the artifacts of core content areas, serving to leverage both notoriety and consistency within those areas. These are two forces from which technology education could certainly benefit.

At the same time, technology educators must be encouraged and empowered to take advantage of the many benefits that can result from the use of classroom-based assessments. Whether formal or informal, measures of student learning that are analyzed to better understand *what* learning has occurred and *how* learning has occurred within technology education classrooms are critical to forward progress within the discipline. Technology teacher education programs are uniquely poised to offer a major contribution in this area. Ensuring that each graduating student has a solid understanding of the basics of assessment development, assessment applications, and data analysis is an essential first step. Second, most in-service teachers need a source of graduate-level coursework or professional development on these topics. Third, technology teacher educators are perhaps best positioned to conduct the kinds of research needed to better understand the teaching and learning process within technology education.

THE PURPOSES OF ASSESSMENT WITHIN TECHNOLOGY EDUCATION CONTEXTS

This yearbook addresses a broad spectrum of types of assessments that might be used within the context of technology education courses or programs. Each type of assessment has specific purposes and applications that must be aligned with the intended use of the resulting data. Designing or selecting an assessment must therefore be a purposeful act, and a well-designed assessment system will typically require a variety of assessment types. The following list identifies and briefly defines a number of common assessment types that are referred to throughout this yearbook:

- **Formative Assessments.** Their primary purpose is to collect timely data about student progress so that teachers can adjust instruction and students can modify learning strategies. They are typically more frequent, less formal, and less comprehensive than other types of assessment.
- **Summative Assessments.** Their primary purpose is to collect data at the end of a learning experience to assign a score or grade to students' performance. They typically occur at the end of a unit or course.
- **Declarative vs. Procedural Assessments.** These terms describe the fundamental purpose of two types of test items. Declarative items are designed to test content knowledge (knowledge about), and procedural assessments are designed to test process knowledge (ability to do).

- **Classroom-Based Assessments.** This term is typically applied to assessments that have been developed by the classroom instructor, school, or district for local use. They may serve a formative or a summative purpose.
- **Commercially Prepared Assessments.** This term is applied to assessments that are prepared by nonprofit or for-profit organizations (e.g., textbook companies) for use in classrooms. Examples include the chapter tests found in textbooks and the module tests developed by the publishers of modular lab programs.
- **Large-Scale Assessments.** Their primary purpose is to collect summative data that can be used to characterize (and compare) learning achievement of groups of students. These assessments are typically developed at the state, national, or international level and include assessments such as the Praxis licensure exams, the SAT and other college entrance exams, and broad-scale assessments such as the National Assessment of Educational Progress (NAEP).
- **High-Stakes Assessments.** This term refers to assessments that are accompanied by consequences (sanctions or rewards) for students, teachers, or school districts. They are typically summative, usually large-scale, and often commercially prepared. Examples include tests that must be passed to receive a diploma, assessments that lead to licensure or accreditation, and mandated end-of-course assessments that may result in either merit pay for teachers or district sanctions against programs or schools.

Aligning Assessment Strategies and Types with Purpose

Validity is a common topic in assessment texts, but it is most often referenced in terms of ensuring the validity of individual test items. However, it is the alignment of assessment strategies and types with purpose that ensures the valid use of the assessment data itself.

Assume, for example, that an assessment was needed for the purpose of improving instruction within some aspect of classroom performance (e.g., using CAD software). Many types of assessment could contribute information regarding students' ability to perform CAD functions, but not all types would be equally effective in producing data that would help the teacher improve instruction. If the purpose were to impact instruction with the current group of students, a summative assessment would simply provide data too late. Therefore, a targeted formative assessment would be a better choice. This situation would also involve choosing between declarative or procedural assessment items. Although using CAD software does require declarative knowledge, an assessment of procedural knowledge would more directly address the stated purpose of learning whether instruction in the use of CAD software has been effective. Next, the teacher would need to decide whether a classroom-based or commercially prepared assessment could provide better data regarding the specific instruction provided. In most cases, a classroom-based assessment will provide more targeted and useful data for the purpose of improving instruction. Combining these considerations would result, therefore, in the selection of a formative, procedural, classroom-based assessment as the optimum choice. This would also be the most valid choice, since validity is increased when alignment between the assessment tool, the resulting data, and the stated purpose is strengthened.

Once the assessment strategy is identified and carried out, however, additional work must be done. The assessment data must be analyzed to determine whether instruction has been adequate, and for whom. Information regarding specific functions or tasks must be teased out of the assessment data, and problems identified. If misunderstandings or inabilities to perform exist, re-teaching to address those problems must occur.

DECIDING WHEN TO ASSESS

Assessment is best thought of as an ongoing process, so the question should really be stated, "At any given time, which is the best assessment to use?" Since the timing of an assessment is ultimately driven by its purpose, clarifying that purpose identifies both the type of assessment needed and the best time frame for its use. It is only by broadening our perspective of assessment to include both formative (ongoing) and summative (periodic) activities that we begin to understand fully the role that assessment plays in the educational process and our responsibility as educators to facilitate that role.

DETERMINING WHAT TO ASSESS IN TECHNOLOGY EDUCATION

Quality assessments can be time-consuming to develop and implement. They should, therefore, produce data that are valuable and that have practical applications within the technology education classroom. The primary applications addressed within this yearbook focus on providing information about and for students, determining teacher/program effectiveness, and measuring the degree to which standards have been met.

Assessment of individual students is done for a variety of purposes. These might include assessing student knowledge of safety within a laboratory environment to reduce the number of accidents; assessing specific units of instruction formatively to show students and teachers needed areas for improvement; or assessing end-of-unit or end-of-course learning attainment for individual accountability and reporting.

Assessing for teacher effectiveness also takes a variety of forms. At the preservice level, assessments can determine student readiness to enter the teaching profession. At the in-service level, assessments can be used to uncover gaps in teachers' ability to reach certain student populations or to achieve course outcomes. Ultimately, assessment of teacher effectiveness could conceivably be used in performance appraisal systems or compensation systems. Similarly, assessments of program effectiveness can be used to determine current program weaknesses and strengths, and to provide insights for planning future changes to the program.

Teaching to standards is increasingly seen as educational best practice. While a variety of standards will find their way into technology education programs, primary among them at this point in time are the *Standards for Technological Literacy*. Important as these standards are for defining the content for the study of technology, they fall short of defining what it means to be technologically literate. According to Pellegrino (2002), content standards cannot substitute for domain-based theories of learning in or about technology. In other words, the standards serve as a starting point for the dialogue about what constitutes knowledge within the discipline, but they must be fleshed out to give a true picture of what this knowledge looks like in context. The questions on which assessment efforts could focus include, "What constitutes expertise in the technology domain?" "What patterns exist in the growth of understanding and competency?" "What constitutes evidence of transfer in technology education?" (Pellegrino, 2002, p. 119). Assessments designed to address questions such as these must necessarily go beyond recall of facts or the ability to mimic the steps in a technological process.

Other foci for assessments include measuring the effects of participation in technology education courses on understanding of "core" knowledge in areas such as mathematics and science; performing needs assessments for programs and staff development; determining customer satisfaction (students, parents, and community); and demonstrating accountability at the local, state, national, and international levels. Furthermore, a variety of political implications that impact what is assessed exists, including marketing of schools and programs as well as influencing policy decisions related to funding, graduation requirements, and international competitiveness.

MAKING ASSESSMENTS COUNT IN WAYS THAT MATTER

It seems that the only constant in the world of education is change, which is typically approached under the auspices of improving education. Too often, the educational tinkering that results is done with little regard to an evidence base (Coalition for Evidence-Based Policy, 2002). Educators may not agree with the politically charged nature of many educational initiatives, or with the rationales sometimes used to promote them. In spite of such agendas, educators cannot lose sight of the fact that assessment data provide the only real proof that change has actually resulted in improvements (or declines) in learning. Furthermore, assessment is a critical step in identifying what needs to change and what types of change might lead to desired improvements. Although assessment in some form has always occurred in educational settings, it has often been underutilized or misused. This is particularly true in the field of technology education and other "noncore" content areas, which have not benefited from the positive effects that can flow from the use of large-scale assessments. Regardless of the level or type of assessment being considered, it is incumbent on all educators to use assessments in ways that improve the educational process.

The Challenge of Creating Quality Assessments

In the world of computer programming, the acronym GIGO indicates that the quality of the output is no better than the quality of the input (garbage in, garbage out). The same lesson applies to assessment: The quality of the resulting data relates directly to the quality of the work done in aligning assessments with purpose, and creating quality items or tasks. The use of quality assessments must also be followed by quality data analysis if maximum benefit is to be gained. A failure in any of these aspects could result in what Popham (2001) called educationally destructive assessments. Furthermore, even a good test can result in bad conclusions: "Most standardized tests ignore the process by which students arrive at an answer, so a miss is as good as a mile and a minor calculation error is interchangeable with a major failure of reasoning" (Kohn, 2000, p. 8).

The problem is exacerbated by the fact that, although tremendous energy goes into devising and administering large-scale assessments, schools and teachers are often not given the resources to follow through on the information gained from these assessments (Stiggins, 2004). Lezotte (1997) notes that districts are given analytical tools to carefully manage their budgets, but that no such tools exist for managing their greatest challenge: monitoring the learning process and intervening to improve that process. Districts are further challenged by the increased emphasis on use of high-stakes testing, which almost inevitably serves to narrow the curriculum by making teachers feel they have less control over what they teach, how they teach it, and how it is assessed (Brennan, 2006). In these circumstances, valuable instructional time is less likely to be used for formative assessments.

These difficulties do not mean that technology educators should abandon the challenge of measuring—in more sophisticated ways—what and how students learn in the technology classroom. This will first require attempting to achieve a more complete and nuanced picture of what it means to be technologically literate, a path of inquiry being traveled by a number of people. At the same time, technology educators must investigate strategies and tools that hold potential for enhancing our ability to track and measure student progress within technology education programs.

The Potential for Computer-Based Assessments

One way to achieve advancements in how we assess students is through the use of computer-based testing tools. At this time, we are just beginning to explore the capabilities presented by computerization in the classroom. Just as with instructional materials, computer-based assessments are often just jazzed-up versions of traditional delivery methods. If we more fully explore the potential of computer-based assessments, we may experience real breakthroughs. For example, audio and video could be incorporated to enhance the way a test is delivered to diverse student audiences. Simulations could be used to enhance the authenticity and/or cognitive demands of a test. Tests could be divided into discrete chunks of information, tailored to specific settings, and delivered on demand to individual students. In this way, checks for understanding could be tightly linked to instruction and scored immediately, making the data available in time for teachers to make adjustments to instruction (Bennett, 2002).

Commercial vendors who develop and market modular systems have played an important role in determining the look and feel of technology education classrooms. The computer-based instructional management tools that are part of many of these systems could be exploited for their potential to transform assessment. These management systems typically allow teachers to track whether and when students have completed tasks, and how many attempts were required to do so. More sophisticated analyses could be applied to these data as a source of information about how students learn (Hoepfl, 2003).

The Potential for Performance Assessments

The literature devoted to promoting the use of performance (or authentic or alternative) assessments is vast. Certainly, within technology education programs it is difficult to accept assessment systems that do not include measures of students' ability to perform tasks that are integral to "doing" technology. It is probably even safe to say that technology teachers have done a credible job of evaluating student performance, in spite of the fact that they sometimes fall prey to a tendency to include nonacademic criteria in their scoring systems (Marzano, 2000). An example might be to include scores for "attitude" or "cleaning the work space" in an evaluation of a design mock-up. Technology teachers can always work to refine performance-scoring tools so that they are more descriptive of the kinds of performance desired, and suggestions for doing so are provided in chapters 4 and 12 of this yearbook. Performance assessments have the potential to provide evidence of learning and skill attainment that could never be captured through a traditional test, and they can be applied in both formative and summative ways.

Although performance assessment is a relatively straightforward process at the classroom level, significant challenges occur when educators attempt to carry them out on a large scale. More specifically, challenges occur when performance assessments are used for purposes of accountability, which seems to demand uniformity of tasks and external scoring to avoid biases. The question is a simple one: What kinds of assessment data will decision-makers accept as evidence of attainment? In Chapter 4, Hoepfl describes a performance assessment pilot study undertaken in North Carolina, and in Chapter 10 Kimbell describes the nationwide performance assessment of design and technology students used in the United Kingdom. Each of these efforts provides some insights into the potential broader use of large-scale performance assessments.

The Potential for Data Analysis Strategies

Several chapters within this yearbook describe analysis strategies that can be used to elicit more sophisticated and useful information from assessment data. For example, in Chapter 3 Haynie describes techniques for measuring item difficulty and item discrimination. These are simple mathematical calculations whose use by technology teachers on a routine basis would result in better traditional test items. Chapter 6 is devoted to data analysis tools and strategies that enable schools and districts to better understand the information gained from testing programs. Perhaps most important, Chapter 6 includes strategies for examining growth in student knowledge over time, as opposed to only taking one-shot measures that often say little about the effectiveness of instruction.

At another level of analysis, in Chapter 11 Ruiz-Primo discusses use of the "assessment square" to evaluate and characterize the nature and comprehensiveness of assessments. This type of analysis is critical for two reasons: It will help to achieve the kind of coherent alignment between curriculum, instruction, and assessment called for by many education experts; and it will provide a means of determining systematically the extent to which standards are being addressed.

These are just a few examples of the benefits that could be attained by adopting better data analysis tools. Clearly, creating adequate measurement devices is only one step on the road toward better assessments. Development of assessments with an eye toward analysis as well as toward tight alignment with curriculum and instruction presents both a challenge and an opportunity for the technology education community.

CONCLUSION

Educational assessment (and use of the associated data) has become a critical national focus and certainly represents a topic whose time has come for technology education. We hope this yearbook promotes what needs to be a continuing dialogue regarding assessment within technology education classrooms and programs. Among future steps, technology educators can become more engaged in the discussion about what constitutes technological literacy at the elementary, secondary, and postsecondary levels. They can become better informed about the issues surrounding, and processes for ensuring, high-quality assessments. They can incorporate new and better assessments into their own work as teachers and program leaders. Finally, technology educators can become better consumers of assessment data, so that better decisions can be made about the future of technology education. As the *Standards for Technological Literacy* are more widely adopted, the technology education profession needs to mature in its use of assessments, both at the classroom/district level and at broader levels. That process should include mechanisms for developing, evaluating, and sharing assessments between systems to enhance the quality and effectiveness of those assessments. We hope this yearbook makes a useful contribution to that dialogue.

DISCUSSION QUESTIONS

1. What do you perceive as the biggest benefits of, and the major problems with, high-stakes assessment in the context of accountability?
2. As a teacher, how do you or could you make use of formative assessments? If you currently teach, how could you change the way you use formative assessments to better inform teaching and learning within your technology education classroom?

3. What emerging assessment strategies or tools do you think hold the greatest potential for improving the way we assess students in technology education? Why?

REFERENCES

Bennett, R. (2002). Using electronic assessment to measure student performance: Online testing. *State Education Standard* 3(3): 23–29.

Black, P., C. Harrison, C., Lee, B. Marshall, and D. Wiliam (2004). Working inside the black box: Assessment for learning in the classroom. *Phi Delta Kappan* 86(1): 8–21.

Brennan, R. L. (2006). Perspectives on the evolution and future of educational measurement. In *Educational measurement,* edited by R. L. Brennan, 1–16. 4th ed. Westport, CT: Praeger Publishers.

Coalition for Evidence-Based Policy (November 2002). *Bringing evidence-driven progress to education: A recommended strategy for the U.S. Department of Education.* http://coexgov.securesites.net/admin/FormManager/filesuploading/coalitionFinRpt.pdf (accessed June 11, 2006).

Hoepfl, M. (2003). Assessment and accountability: Where are we headed? Paper presented at the 90th Annual Mississippi Valley Technology Teacher Education Conference, Nashville, TN.

International Technology Education Association (2000). *Standards for technological literacy: Content for the study of technology.* Reston, VA: International Technology Education Association.

——— (2003). *Advancing excellence in technological literacy: Student assessment, professional development, and program standards.* Reston, VA: International Technology Education Association.

——— (2004). *Measuring progress: Assessing students for technological literacy.* Reston, VA: International Technology Education Association.

Kohn, A. (2000). *The case against standardized testing: Raising the scores, ruining the schools.* Portsmouth, NH: Heinemann.

Lezotte, L. W. (1997). *Learning for all.* Okemos, MI: Effective Schools Products Ltd.

Marzano, R. J. (2000). *Transforming classroom grading.* Alexandria, VA: Association for Supervision and Curriculum Development.

National Academy of Engineering (NAE) and National Research Council (NRC) (2006). *Tech tally: approaches to assessing technological literacy.* Edited by E. Garmire and G. Pearson. Washington, DC: National Academies Press.

Pellegrino, J. (2002). How people learn: Contributions to framing a research agenda for technology education. In *Learning in technology education: Challenges for the 21st century,* edited by H. Middleton, M. Pavlova, and D. Roebuck, 114–129. Vol. 2. Brisbane, Australia: Griffith University Centre for Technology Education Research.

Pellegrino, J., N. Chudowsky, and R. Glaser (2001). *Knowing what students know: The science and design of educational assessment.* Washington, DC: National Academies Press.

Popham, W. J. (2001). *The truth about testing: An educator's call to action.* Alexandria, VA: Association for Supervision and Curriculum Development.

Stiggins, R. (2004). New assessment beliefs for a new school mission. *Phi Delta Kappan* 86(1): 22–27.

Assessing Technological Literacy: A National Academies Perspective

Chapter 2

Rodney L. Custer
Illinois State University

Greg Pearson
National Academy of Engineering

INTRODUCTION

The National Academies (www.nationalacademies.org), based in Washington, D.C., is a nonprofit, nongovernmental set of institutions that has a long history of addressing issues at the intersection of science, engineering, and public policy. One topic that has received considerable attention from the Academies, particularly the National Academy of Engineering (NAE), is technological literacy. In late 2005, the Academies completed a two-year study that focused on determining approaches for assessing how much students, teachers, and out-of-school adults know and can do with respect to technology. This chapter addresses many of the concerns in the study and discusses some of the findings and recommendations in the NAE's report.

The National Academies

Founded in the 1860s as the National Academy of Sciences, the Academies have grown to include the National Research Council (NRC), the National Academy of Engineering (NAE), and the Institute of Medicine. In addition to serving as honorary membership organizations, these institutions conduct hundreds of studies every year, often at the request of Congress or the executive branch, on a variety of topics. Committees of experts who serve without compensation oversee the studies, which may be funded by the federal government, corporate or nonprofit foundations, individuals, or internal monies. In a typical year, there are some 10,000 volunteer experts and 1,200 staff working on nearly 700 studies. Annually, this work results in about 250 published reports.

The process of selecting individuals to serve on study committees includes steps that assure the panel has all of the requisite expertise, that there is a balance among panelists' points of view on the issue at hand, and that there is no conflict of interest that might impugn the study results. An assessment report is reviewed by a group of 11 individuals not directly involved in the project, but who have expertise very similar to those on the study committee. The identities of the reviewers are not revealed to the committee or project staff until after the report has been approved for publication. Reviewers' comments result in a number of editorial and organizational changes that improve the accuracy, completeness, and readability of the final document.

The National Academy of Engineering

The NAE's involvement in technological literacy began in the mid-1990s with discussions between the then interim NAE President William Wulf and a number of individuals from inside and outside the Academies. Wulf was interested in involving the NAE in K–12 education issues, a passion that he brought with him from the University of Virginia, where he is a professor in computer science. One important early contact was with Kendall Starkweather, executive director of the International Technology Education Association (ITEA). At the time, ITEA was just beginning its effort to create K–12 content standards. In a move that would affect not only the standards, but also the course of technology education in the United States, ITEA received input from the NAE and NRC on developing standards. The NAE soon adopted the "technological literacy" label to describe its portfolio of studies, workshops, and other activities intended to promote and improve public understanding of technology.

The National Science Foundation

In the late 1990s, the National Science Foundation (NSF) provided funds for a joint NAE-NRC project that was meant to draw attention to and make the case for technological literacy. The resulting report, *Technically Speaking: Why All Americans Need to Know More About Technology* (National Academy of Engineering and National Research Council, 2002), argued that all citizens ought to have some level of technological knowledge and capability. The project determined that there had been very few attempts to actually measure technological literacy in adults or children. The report noted that without a way to ascertain an individual's or a population's technological literacy, it is not possible to determine whether steps to enhance such literacy are working. To address this concern, *Technically Speaking* recommended an initiative to assess technological literacy.

In 2002, again with funding from NSF, NAE partnered with the NRC's Center for Education to analyze the need, extent, and possible approaches for assessing technological literacy in the United States. A 16-member study committee, chaired by NAE member Elsa Garmire of Dartmouth College, included four technology educators: Rodney Custer, then professor and chairperson of the Department of Technology at Illinois State University; Marc DeVries, an assistant professor of philosophy and methodology of technology at Eindhoven University of Technology in the Netherlands; William Dugger, director of the Technology for All Americans Project in Blacksburg, Virginia; and Richard Kimbell, head of the Technology Education Research Unit at Goldsmiths College, University of London. Individuals with expertise in assessment design and implementation, K–12 science and mathematics education, curriculum and standards development, informal education, cognitive science, public understanding of science and technology, simulation, and sociology were also on the committee.

The committee's report, *Tech Tally: Approaches to Assessing Technological Literacy,* was published in summer 2006. This chapter discusses several key aspects of the report, including a conceptual framework for assessing technological literacy in students, teachers, and out-of-school adults; the challenge of assessing technological capability; and recommendations for improving and expanding assessment in this arena.

THE CASE FOR ASSESSING TECHNOLOGICAL LITERACY

It is clear that we are living in a world that has become increasingly technological. For nearly everyone, part of living in the 21st century involves having at least some technical knowledge and capability. This is certain to increase in the years and decades to come:

> As far into the future as our imaginations can take us, we will face challenges that depend on the development and application of technology. To take full advantage of the benefits and to recognize, address, or even avoid the pitfalls of technology, Americans must become better stewards of technological change. (NAE and NRC, 2002, p. 12)

Five compelling arguments are presented in the NAE and NRC publication *Technically Speaking* that support the value of technological literacy: (1) making good consumer decisions; (2) increasing citizen participation and involvement in the democratic process (i.e., voting on technologically related issues); (3) helping individuals develop a base level of technological skill as a foundation for workforce development; (4) ensuring access by narrowing the digital divide; and (5) enhancing social well-being by providing citizens with a sense of empowerment with, and over, technology.

The potential benefits of assessing technological literacy may be significant, but that fact by itself will not bring such assessments into being. Any discussion of assessment in the United States today must consider the overall climate for educational testing and accountability, particularly related to the No Child Left Behind Act (NCLB). NCLB has placed considerable pressure on schools to demonstrate achievement gains in reading, mathematics, and, beginning in 2008, in science for students in grades three through eight and at least one year of high school. The law also requires all K–12 teachers to be highly qualified in the subjects they teach. Finally, NCLB asks that states assure every student is technologically literate by the eighth grade. Although the law does not define technological literacy, the requirement has been largely interpreted by the states to refer to computer, or information, technology literacy.

In this environment, it is easy to see how any initiative that increases the overall assessment burden might fail to gain traction. Compared with more established areas of testing—i.e., reading literacy and numeracy—assessment for technological literacy will face an additional challenge: It does not have a strong connection to a core K–12 school subject. Technology education is the academic base from which assessment of technological literacy ought to spring. But the field's small size and historical roots in industrial arts education, among other factors, continue to challenge efforts to bring it into the educational mainstream. Unless or until there is greater emphasis in K–12 classrooms on curricula that encourage the study of technology, it is hard to imagine widespread, school-based assessment of technological literacy.

However, the situation is far from hopeless. As important as it is for technology education to continue to grow and become more relevant in the broad sweep of American education, technology education is not the only discipline with a claim on technological literacy. Science, mathematics, social studies, and, increasingly, engineering are K–12 subjects that can include explicit content connections to technology. Assessments in these subjects might also—and in some cases already do—include technology-related items.

Compared with the situation for K–12 students and teachers, assessing out-of-school adults presents different challenges. For one thing, adults are not the captive audience that children or teachers sitting in a classroom are, so anyone assessing adults will need to cope with the logistics of identifying appropriate test subjects and convincing them to participate. In addition, assessing adult populations is not required by the government the way it is for K–12 students, so those interested in assessing what adults know about technology will have to raise funds to carry out such efforts.

A CONCEPTUAL FRAMEWORK FOR ASSESSING TECHNOLOGICAL LITERACY

To assess technological literacy, conceptual frameworks are valuable tools with which to structure complex ideas into a format that can be easily understood and communicated with a variety of constituencies. Conceptual frameworks are frequently presented in graphic form, often as a matrix. These graphic structures typically identify the essential elements of a concept and then attempt to describe how these elements interact or correspond with one another.

The value of conceptual frameworks extends far beyond identifying, clarifying, and structuring ideas. These frameworks are essential to assessment planning. Assessment designers rely heavily on these conceptual structures, using them as blueprints to ensure comprehensive coverage of all aspects of a topic, concept, or field of study.

The National Assessment Governing Board

A leading authority in education assessment is the National Assessment Governing Board (NAGB), which since 1969 has overseen the National Assessment of Educational Progress (NAEP). The NAEP, widely known as the Nation's Report Card, is designed to determine what U.S. students know and can do in various subject areas. NAEP assessments are designed with the help of detailed frameworks and associated matrices that contain cognitive and content elements. For example, the "cognitive dimension" of the NAEP science framework (NAGB, 2004) includes conceptual understanding, scientific investigating, and practical reasoning; whereas the "content dimension" includes earth science, physical science, and life systems. The nature of science and themes related to systems, models, and patterns of change are presented as cross-cutting "big ideas." The current science framework, developed in 1990, is being revised to use in the 2009 science assessment (NAGB, 2004). The NAGB has developed similar cognitive/content frameworks to structure assessment planning for a broad range of academic disciplines in addition to science, including reading, mathematics, geography, U.S. history, and civics.

The Conceptual Framework of Assessment

Several documents published in the past decade provide starting points for those interested in developing a detailed conceptual framework to assess technological literacy. The *Standards for Technological Literacy* (ITEA, 2000), its companion document *Advancing Excellence in Technological Literacy* (ITEA, 2003), and *Technically Speaking: Why All Americans Need to Know More About Technology* (NAE and NRC, 2002) present schemes for organizing the content of technology. In addition, standards developed for science by the American Association for the Advancement of Science (AAAS, 1993) and the National Research Council (NRC, 1996) include significant technology-related elements.

Tech Tally (NAE and NRC, 2006) presents a two-dimensional matrix that combines core organizing ideas from the *Standards for Technological Literacy* and *Technically Speaking*. *Technically Speaking's* three dimensions of technological literacy (knowledge, capability, and ways of thinking and acting) identify a broad spectrum of cognitive activity and are conceptually consistent with the cognitive elements in NAEP's science and mathematics frameworks. As a comprehensive set of content standards, the ITEA's *Standards for Technological Literacy* (STL) represents the logical point of departure for structuring the content dimension.

Through careful deliberation, the study committee made some minor modifications to the STL's approach, combining the "understanding" and "doing" of design into a single content dimension, and merging the Designed World standards into a content dimension, entitled "Products and Systems."

The final content schema consists of characteristics, core concepts, and connections that closely align with the STL. The resulting conceptual framework has the advantage of forming ranks with contemporary national assessment planning while retaining the essential content elements identified in key curriculum standards documents (Fig. 1).

Figure 1. Conceptual framework for technological literacy.
(Source: NAE and NRC, 2006, p. 53).

Cognitive Dimensions

	Knowledge	Capabilities	Ways of Thinking and Acting
Technology & Society			
Design			
Products & Systems			
Characteristics, Core Concepts & Connections			

Conceptual understanding will continue to change and evolve over time. This has certainly been the case with technological literacy as a concept as well as with the field of technology education, which has redefined itself substantially over the past two decades. For the purpose of assessment, a conceptual framework, such as the one described in this section, represents a structure that aligns with multidisciplinary, national-level practice (NAEP) and provides a healthy environment within which understanding technological literacy can continue to evolve and be refined.

TECHNOLOGICAL LITERACY AND THE NATIONAL ASSESSMENT AGENDA

For the purpose of influencing national- and state-level changes in education policy and practice, data from large-scale assessments are crucial. Large-scale student assessments typically cross classroom and school boundaries and provide information to schools, school districts, and states about trends in student achievement. The results of such assessments also provide accountability. This is the case with NCLB, discussed earlier, which requires states to meet annual goals for improving student performance in certain core subjects. These types of state-level assessments reveal the extent to which individual students are meeting subject- and grade-specific learning goals such as those spelled out in curriculum frameworks.

At the national level, it is not feasible to have each individual answer every question, so large-scale assessments such as NAEP are based on sampling techniques. Under this method, subgroups of test subjects are given only a portion of the total battery of test items. Because no single test taker answers every question, comprehensive scores cannot be computed for individual students, and sample-based results are not used for accountability purposes. However, measures of performance for the group as a whole—say, fifth graders—can be computed. Therefore, such assessments are helpful for tracking national trends in performance according to various demographic variables, such as gender, race, and socioeconomic status. In addition, in the case of NAEP, sample-based performance data for states are also reported. For both state and national samples, NAEP converts the raw scale scores into three broad achievement levels: "basic," "proficient," and "advanced." Criteria for each level are developed with input from subject-matter experts, parents, and others. It is worth reminding readers here that technological literacy is not currently part of the assessment agenda at the national level in the United States. However, some states (including North Carolina—see Chapter 8) do conduct statewide standardized testing of all students enrolled in technology education courses.

The Connection to Government Testing

How might assessment for technological literacy fit within these existing state and national assessment regimes? There are at least two possibilities. The first would be to include test items related to technology in existing assessments for other subjects, such as science, mathematics, and history. The natural subject connection between technology and these other fields makes this approach intuitively attractive. At least one researcher has made the case that NAEP's long-term science assessment included items consistent with a vision for technological literacy (Hatch, 1985, 2004). This approach is limited by the small number of included items that makes it difficult to draw statistically valid conclusions about the technological literacy of an individual or a group.

A second possibility would be to craft an entirely new assessment instrument devoted to probing the multiple aspects of technological literacy. While certainly more costly and time-consuming, such an assessment would have obvious advantages. For one thing, it would produce more valid and reliable performance data than the first approach, potentially across all three dimensions of technological literacy. It would also provide a highly visible indicator of U.S. students' technological savvy and bring attention to the field of technology education. Depending on the nature of the assessment results, it could also stimulate policy makers to enhance technological-literacy-related curriculum development, teacher education, and research.

In this regard, one important development is NAGB's 2005 decision to conduct a "probe study" of technological literacy assessment. The board uses probe studies to test the feasibility of designing new stand-alone assessments. The study will have several phases, which include developing an assessment framework, field-testing of sample items, and assessing students, which will occur in 2012. The rollout and results of the probe study, which will be made public, will bear the scrutiny of those in the technology education community.

THE SPECIAL CASE OF PERFORMANCE ASSESSMENT

In *Technically Speaking*, technological literacy has three dimensions: knowledge, capability, and ways of thinking and acting. The capability to do something with a variety of tools and technologies, as well as to solve technological challenges, has been at the core of the technology education field throughout its history. In fact, this historic attention to capability has frequently led to a disproportionate emphasis on activities (often for their own sake) with a corresponding de-emphasis on content and conceptual development (Custer, 2003). While the STL have identified and elevated the importance of technological concepts and critical thinking, the ability to "do" will remain essential to technology education.

The Challenge of Assessing Capability

This emphasis on capability presents some unique challenges for assessing technological literacy. First, the general public tends to associate the capability of a technologically literate individual with the in-depth technical competence of specialized technicians and technologists (e.g., electricians, automotive technicians, computer repair specialists, etc.). A clear distinction should be drawn between these two different skill sets. To broadly assess technological literacy, the focus should clearly remain on more general levels of capability. Based on the STL, these include capabilities such as applying general design and problem-solving processes, using basic tools for practical purposes, and performing simple repairs on household devices. One significant ongoing challenge for technology educators will be to clarify the specific types and levels of capability appropriate for technological literacy.

A second challenge for assessing technological capability has to do with aligning capability with conceptual development. As noted in *Technically Speaking*, technological literacy represents a combination of knowing, doing, and thinking critically in relation to technology. The STL represents a concerted effort to identify the core concepts connected with the activities that are being delivered in technology education laboratories. While the alignment between concept and capability is important and necessary, it poses significant challenges for assessment. Developing assessments that are able to tease conceptual knowledge out of activity-based curricula will not be easy. Current assessment practices tend to separate the two, focusing on either knowledge or performance.

A third major challenge for assessing the capability dimension of technology literacy involves large-scale assessment. Because of the resources and time required to test large numbers of individuals, large-scale assessments typically include mostly multiple-choice and short-answer formats, since these can be scored quickly and easily. Performance assessments, on the other hand, tend to be highly individualized and typically involve direct student observation and other time- and resource-intensive practices. Furthermore, specialized knowledge, expertise, and training are often required for evaluators to objectify student performance into norms that could apply to selected populations. One notable example of successful, national-level, large-scale assessment was developed and conducted in the United Kingdom in the late 1980s by a team of researchers led by Richard Kimbell. *The Assessment of Performance in Design and Technology* project assessed the design ability of 10,000 15-year-old students using a paper-and-pencil instrument and focused on the designing aspects of technological capability. The rubric-based assessment involved extensive training for the rater and was implemented on a national scale.

Simulated Environments for Assessment

Some of the challenges associated with performance assessment may be addressed with tools such as computer-based simulations. An entire chapter of *Tech Tally* is devoted to this topic. Simulated environments offer the possibility of assessing performance and competence in situations that cannot or should not be attempted in actual practice. In education, simulations and computer-based modeling are being investigated as tools to enhance learning in areas such as biology, chemistry, and physics (see, for example, projects of The Concord Consortium [www.concord.org] and the Technology Enhanced Learning in Science center [www.telscenter.org]). Web-based electronic portfolios offer another avenue for assessing students' design-related capabilities. The United Kingdom is developing an e-portfolio system (e-scape) that contains solutions to creatively assess, in e-portfolio environments, work done by students in design and technology classes. The system, in which students use personal digital assistants—such as sketchbooks, notebooks, cameras, and voice recorders—was pilot tested in 15 schools in summer 2006 (R. Kimbell, personal communication, August 17, 2006).

RECOMMENDATIONS FOR IMPROVING ASSESSMENT OF TECHNOLOGICAL LITERACY

Tech Tally makes 12 recommendations intended to improve and expand the assessment of technological literacy in the three target populations: K–12 students, K–12 teachers, and out-of-school adults. In this section, we will address only the recommendations focused on teachers.

The report argues that all K–12 teachers, not just technology educators, need a basic level of technological literacy. Technology is integral to many, if not most, school subjects—from history to art to science—so teachers should be comfortable discussing technology-related issues. Put another way, can we expect students to be fully technologically literate unless their teachers are? Yet apart from the Educational Testing Service's Praxis exam for technology teachers and the technology portion of the Praxis exam for science teachers, there are no assessments aimed at elucidating teachers' technological understanding. Even the Praxis assessments, in the study committee's view, fell short of adequately probing the knowledge, capabilities, and critical thinking spelled out in such documents as *Technically Speaking* and the ITEA *Standards*.

Tech Tally makes two recommendations related to teacher assessments. First, the report suggests that questions about technology be included in the tests that states use to determine teacher quality under provisions of NCLB. Since avenues other than tests can be used by teachers to meet the NCLB quality standard, this recommendation, if implemented, can be expected to have only moderate impact. Second, the report urges the National Science Foundation and the U.S. Department of Education to fund the development of stand-alone, sample-based assessments of pre-service and in-service teachers of science, technology, English, social studies, and mathematics. Such wide-scale testing could have significant impact on how schools of education prepare teachers, on whether and how prominently technology is included in K–12 curricula, and on the activities of state boards of education and other groups concerned with teacher preparation and teacher quality.

IMPLICATIONS FOR THE EDUCATION OF TECHNOLOGY TEACHERS

The focus and scope of this chapter has extended beyond the field of technology education. This is important and intentional since technological literacy and its assessment are, to some extent, embedded into the content of other academic disciplines and its implications impact society. Another key indicator of this broad perspective is the fact that technological literacy has captured the attention of prestigious organizations such as the National Academies, NSF, and NAGB.

Technology teacher educators should pay careful attention to this increased visibility and expanded context. As the technology education field continues to evolve, teachers and teacher educators will increasingly need to know about, and be involved with, cross-disciplinary efforts to promote, deliver, and assess technological literacy. This should include familiarity with various standards documents, national assessment organizations and initiatives, and the work of state and national agencies and educational institutions. This is important for curricular and advocacy reasons, and will also equip technology educators to collaborate meaningfully with educators from other subject areas, such as mathematics, science, and engineering, to conceptualize, deliver, and assess technological literacy. The collective goal should be to position technological literacy on the national assessment and policy-making agenda. Such positioning is critical if addressing technological literacy is to become a part of what is considered a core function of schools across the United States.

DISCUSSION QUESTIONS

1. The *Tech Tally* report recommends developing a "stand-alone, sample-based assessment of pre-service teachers of science, technology, English, social studies, and mathematics" (2006, p. 9). Based on the materials presented in this chapter and standards from the disciplines, what types of knowledge, capabilities, critical thinking, and decision making specific to technology do you think should be assessed?

2. *Tech Tally* alludes to the special challenges associated with assessing performance, particularly on a large-scale basis. What would be required to develop a successful assessment effort of this type, and what do you think its impact would be outside of technology education?
3. It has been said that we value what we assess. Based on your knowledge of assessment, what do you think it would take to convince policy makers, educators, and other key constituencies to develop and deploy assessments of technological literacy?

REFERENCES

American Association for the Advancement of Science (AAAS) (1993). Benchmarks for science literacy. *Project 2061.* New York: Oxford University Press.

Cross, N. G. (2000). *Engineering design methods: Strategies for product design.* New York: Wiley.

Custer, R. L. (2003). *Technology education in the U.S.: A status report. In Initiatives in technology education: Comparative perspectives,* edited by G. Martin and H. Middleton. Griffith University: Technical Foundation of America.

Hatch, L. O. (1985). Technological literacy: A secondary analysis of the NAEP science data. PhD diss., University of Maryland, College Park.

Hatch, L. O. (2004). Technological literacy: Trends in academic progress: A secondary analysis of the NAEP long-term science data. Paper commissioned by the National Research Council Committee on Assessing Technological Literacy.

International Technology Education Association (2000). *Standards for technological literacy: Content for the study of technology.* Reston, VA: International Technology Education Association.

——— (2003). *Advancing excellence in technological literacy: Student assessment, professional development, and program standards.* Reston, VA: International Technology Education Association.

Kimbell, R. K., T. Stables, A. Wheeler, A. Wosniak, and V. Kelly (1991). *The assessment of performance in design and technology: The final report of the APU design and technology project 1985–1991.* London: School Examinations and Assessment Council.

National Academy of Engineering (NAE) and National Research Council (NRC) (2002). *Technically speaking: Why all Americans need to know more about technology.* Edited by G. Pearson and T. Young. Washington, DC: National Academies Press.

——— (2006). *Tech tally: approaches to assessing technological literacy.* Edited by E. Garmire and G. Pearson. Washington, DC: National Academies Press.

National Assessment Governing Board (NAGB) (1996). *Mathematics framework for the 2003 National Assessment of Educational Progress.* http://www.nagb.org/pubs/math_framework/toc.html (accessed August 31, 2004).

——— (2004). *Science framework for the 2005 National Assessment of Educational Progress.* http://nagb.org/pubs/s_framework_05/toc.html (accessed September 8, 2005).

——— (2005). *Science NAEP 2009: Science framework for the 2009 National Assessment of Educational Progress-prepublication edition (forthcoming).* Developed by WestEd and the Council of Chief State School Officers under contract to the National Assessment Governing Board. http://www.nagb.org/pubs/naep_fw_pre_pub_edition_for_web.doc (accessed February 15, 2006).

National Research Council (NRC) (1996). National science education standards. *Standards and benchmarks for science and technology.* Washington, DC: National Academies Press.

Conventional Classroom Assessment Tools: Item Selection, Development, and Evaluation

Chapter 3

W. J. Haynie III
North Carolina State University

INTRODUCTION

Technology education, from its earliest roots in apprenticeship, industrial education, Sloyd, manual arts, and industrial arts, has always been a "learn-by-doing" discipline. The reader may therefore ask what value traditional classroom tests have in this hands-on approach to learning. It is clearly apparent that many of the skills students learn in technology education classes are better evaluated via the products of their experiences, observation of their efforts, rubrics, self-assessments, and other means, but cognitive knowledge is still often best evaluated by tests. This chapter will help you understand how to prepare and administer effective tests appropriate for technology education.

In general, teachers demonstrate a lack of skill in development of effective tests (Burdin, 1982; Carter, 1984; Fleming and Chambers, 1983; Gullickson and Ellwein, 1985; Stiggins and Bridgeford, 1985). Two studies by Haynie (1992; 1997a) carefully examined teacher-made tests from technology classrooms to see if the same could be said in our discipline. Those findings clearly support the need for a chapter in this yearbook concerning test development. Teacher-made tests are especially important because of their unique ability to be tailored to specific instructional objectives and to match the learning conditions in a given classroom (Mehrens and Lehmann, 1987; Nitko, 1989).

Therefore, the basic goal of this chapter is to help technology teachers improve the quality and appropriateness of tests they develop. Since many technology teachers also use commercially developed tests and test items (found in textbook teachers' guides and modular instructional units), a secondary goal is to help teachers better evaluate, select, and edit or customize items and tests from those sources. Topics concerning the importance of traditional classroom tests, development of effective tests, administration of tests, analysis of tests for improvement, and the limitations and pitfalls of traditional classroom tests are presented. Teachers who heed the recommendations here will find their tests more helpful in their efforts to fairly and fully evaluate their students.

IMPORTANCE OF TRADITIONAL CLASSROOM TESTS

Any examination of the *Standards for Technological Literacy* (ITEA, 2000) makes clear that today's technology courses are rich in cognitive content on a greater diversity of topics than ever before. Much of this content knowledge can best be assessed via tests.

Knowledge of information, understanding of concepts, and applications of principles are appropriately evaluated with tests. Likely, the reader recognizes these as the first three levels of Bloom's taxonomy (1956). In many school subjects, tests are used for evaluation of nearly all learning, but in technology education we have historically evaluated higher-order cognitive knowledge (i.e., analysis, synthesis, and evaluation—Bloom, levels 4–6) via projects, presentations, group activities, and other assignments, rather than with tests. Readers should note that throughout the rest of the chapter the terminology adopted in the Revised Bloom's Taxonomy (RBT) will be used to refer to the levels within the cognitive-process dimension: remember, understand, apply, analyze, evaluate, and create (Anderson et al., 2001).

Knowledge and understanding are most often evaluated with tests in all disciplines, including technology education. The application level has multiple aspects, some of which are best evaluated on the basis of projects and longer assignments, but others that are well assessed with classroom tests. Tests can also be good for determining how much a student learned independently from a group activity or large project on which the student may have had considerable help from a parent, other teachers, or peers.

Rationale for the Use of Traditional Tests

One criticism of tests is that students simply memorize facts to pass the tests. The criticism proves valid when tests intended to measure learning at the understanding level or higher levels can be passed via memorization alone. Those tests need items that force students to demonstrate understandings and applications rather than simple facts. Examples of valid applications of testing at the lowest level of RBT (simple recall of facts) would be rote memory of lab safety rules, spelling of technical terms, or a sequence of process steps that must memorized. In these cases, students need "automatic" recall of basic facts so they do not have to look up the information repeatedly, and a quiz or test will encourage them to learn the basics before engaging in more advanced study.

Tests that ask students to restate meanings in their own terms, select the best answer from several choices of which all but one are only partially correct, or apply learned information in a new situation can help teachers determine which students have merely memorized facts without understanding. These questions separate the students who committed information temporarily to short-term memory from those who really learned.

Efficiency is a consideration when deciding to test. A test can quickly demonstrate which students are ready to move to the next step of a process or next unit of study when the actual project they are working on will not be finished until much later. Suppose students in a medical technology unit are assigned to develop a Web page about a medical condition. If the overriding objectives of the unit demand that all students be able to define "diagnosis" and explain how technology affects treatment, it would be conceivable that a student could explain a condition but never truly attain the stated objectives of the unit. The student might know more than anyone in the class about one condition, but nothing in the project would demonstrate adequately that they met the major objectives. However, if a test were given following the student presentations in which all of the students had to apply what they learned (both from their own project and from hearing the presentations of peers) to a novel situation, such as discovery of a new disease and its impacts, then the teacher would have a good basis to assess which students had truly attained the major objectives and which ones simply learned a lot of facts about one disease. Such a test surely would be immune to the criticism of simple memorized facts and would represent an excellent use of classroom testing in a modern technology education laboratory.

Conventional Classroom Assessment Tools: Item Selection, Development, and Evaluation

How and When Traditional Tests Are Used

The above examples show some instances in which tests are used effectively in technology education. They can test knowledge of facts or demonstrate understanding of concepts and relationships. They can be designed to ask students to apply learned information in new settings. They are often the most efficient way of assessing understanding when everything that is important simply cannot be infused into other student assignments and projects, when it is difficult to evaluate what part of a group project an individual student experienced, or any situation in which the test item clearly matches the stated learning objectives. To meet this challenge, however, the test items must be good ones. A poorly developed test will be a hindrance to both learning and evaluation and will waste valuable learning time. Upcoming sections will explain how to develop effective tests, how to ensure fairness and validity, and how to administer tests properly to maximize their usefulness and minimize the stress they produce in students.

DEVELOPING EFFECTIVE CLASSROOM TESTS

All teachers can recall the serious blow to their egos when they graded the very first test they prepared and administered in a class. Despite the typical novice's assumption that students simply lacked ability or effort, most of us later came to realize that our test was probably flawed or that we had not actually found the best approach for teaching the material. Our inexperience may have deluded us into the assumption that the students did know the material well, just because of the lack of questions at the end of each day's lesson when we queried, "Does everyone understand?" With experience, we later understood that students rarely give response to that overused question. So, we believed they did know the material until we graded that first test. What a learning experience it was for us! Even the best students cannot perform well on a poorly developed test.

Types of Test Items

The first choice to make is what type of test to develop. Some types of tests are easy to develop quickly, while others require lots of work up front.

There is, however, a trade-off to consider. The ones that are easy to develop often take longer to grade. Before you take the easy road of simply writing three essay questions on the chalkboard, remember that this approach will have you up late several nights reading all those answers. So, practical concerns of time are to be considered. Table 1 shows a listing of types of test items and some characteristics of each.

Table 1: Types of test items and their characteristics.

Type of item	Advantages	Disadvantages
Essay	Easy to develop, often the best way to allow students to put information into their own words (understanding) or to force them to put their understandings into a new context (application).	Difficult to grade, burdensome for students, and may allow students to go astray and get lost in details that are less important than the essential information you wish to test.
True-False	Few, see text.	Easy to develop poor ones, difficult to develop good ones; easy to grade; often seem tricky or ambiguous to students; more often fraught with errors than any other type of item (Haynie, 1992, 1997a); more likely to be answered incorrectly by the best students, while weaker students prefer them because they seem easy. Generally not recommended.
Multiple-Choice	Easy to grade; can test at any level of RBT; for the skilled test developer, these are the best and most versatile items (but teacher-developed ones often have serious flaws and caution is recommended [Haynie, 1992, 1997a]).	Must be carefully developed to avoid the extremes of ambiguity (due to tricky response alternatives) or simplicity (due to some response choices being so implausible that even uninformed people would never select them).
Matching	Relatively easy to develop and easy to grade.	Only useful when there are lists of related stems and responses which are compatible; often misused by teachers.
Short Answer	Generally easy to develop. If the pitfalls are avoided, these can be very useful for classroom tests. This would also include listing items.	Tend to test at lower levels of cognition; can be ambiguous if not properly developed.
Problems	Especially effective when mathematical computation or use of an algorithm is an important aspect of the topic of study. Word or story problems are better for testing higher level understandings and applications.	More difficult for slower students who are less able to separate the important facts from trivial details.

Matching Tests to Objectives

Before committing to a type of test or test item, the teacher must carefully consider the objective(s) to be tested. An objective written at the application level of RBT cannot be tested with a simple "List the six stages of . . ." sort of item—that would be basic recall of facts. Likewise, if the objective is for students to be able to "list the four basic components of the systems model," an essay item that asks them to explain the operation of a 3-D prototyping machine would be unfair. In fact, a student answering such a question would probe around all four of the components of the systems model (input, process, output, feedback), but may not actually name any of them—thus the teacher would not know whether students giving the best answer had met the specific learning objective intended. So, matching item type to objective level is crucial. Recall knowledge (lowest level) is tested with lists, short answer, true-false, matching, and multiple-choice items. Second level cognitive objectives (understanding) are better tested with brief essays, multiple-choice, and sometimes matching items. Application-level objectives are well tested with problems (especially story problems), essay, and multiple-choice items.

Developing Good Test Items

Care in matching item type to the level of the objective it tests is clear; another related issue is the overall balance of the test. If no purposeful action is taken to ensure balance, what often happens is that items that are easy to develop dominate the test, while some important information may go entirely untested. To prevent this, construct a "test blueprint" (also known as a table of specifications) prior to writing the items (Bertrund and Cebula, 1980; Gronlund, 1981). The test blueprint is a matrix with the objectives to be tested on the left side and multiple columns with headings to show how many and what levels of items are to be developed for each objective (see Table 2). The types of items to develop will naturally follow from the levels of the objectives, as discussed earlier.

Table 2. Portion of a sample test blueprint on GIS technology.

Objectives Students will be able to:	Level 1 Remember	Level 2 Understand	Level 3 Apply	Level 4 Analyze	Level 5 Evaluate	Level 6 Create	Total
List six applications of GIS technology	3						3
Identify and explain how GIS technology improves understanding of resource utilization	1	2	1				4
Apply the principles of GIS technology in a novel simulated setting		1	2				3
Define the key terms common in GIS applications	4						4
Develop a hypothetical scenario in which GIS technology could be applied to solve a societal problem				1			1
Discuss how GIS applications might be used by a state government		3	1				4

Now let's consider some tips on how to develop improved items of each type. Example items are provided and critiqued.

True-False Items

Though they are often plagued with problems, true-false items are popular among teachers, and students are accustomed to them. Teachers who choose to use them must, however, take care to avoid their natural pitfalls. The first consideration is to avoid words such as "never" and "always," which consistently make statements false. Next, ensure that statements intended to be true are indeed fully true, with no possible false or contradictory information imbedded within them. In constructing items intended to be false, it is imperative that the element that renders the statement untrue is an obviously important element of the statement—not something silly or insignificant, such as a misspelled word or unimportant detail tagged into an otherwise true statement. The intention is to determine how much of the subject matter students know, rather than to see which of them is the best detail proofreader, most talented trivia buff, or best able to decipher complex sentences. True-false items often have serious flaws that make them ambiguous (and therefore tricky) in such a way

Conventional Classroom Assessment Tools: Item Selection, Development, and Evaluation

as to cause the brightest students to agonize over them and sometimes mark them incorrectly, when they actually know the subject matter well. At the same time, some of the less-informed students score high by guessing correctly; after all, there is a 50% chance of a right answer. One of the most anxiety-provoking tests this author ever experienced was a college exam that consisted entirely of 100 true-false items. Every item was true. Even the best students in the class labored over this test for the entire period, feeling certain that some chicanery was at hand and that many of the statements were surely false. Some students scored very poorly on the test. Such practices simply cannot be defended.

Figure 1. Sample true-false items.

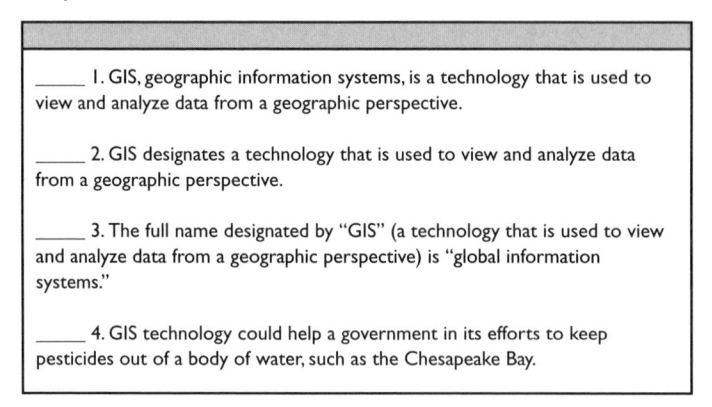

_____ 1. GIS, geographic information systems, is a technology that is used to view and analyze data from a geographic perspective.

_____ 2. GIS designates a technology that is used to view and analyze data from a geographic perspective.

_____ 3. The full name designated by "GIS" (a technology that is used to view and analyze data from a geographic perspective) is "global information systems."

_____ 4. GIS technology could help a government in its efforts to keep pesticides out of a body of water, such as the Chesapeake Bay.

Examining the true-false items in Figure 1 illustrates some of these points. Item 1 is a true statement, with one word (Information) misspelled. If the teacher's intent was to test students' spelling of that word, the item should simply read, "The proper spelling of the middle word designated by GIS is Information." This would not be a particularly important test item, but at least it would not be a "trick question," with its purpose embedded in a distracting field. If the item's actual intent is to determine that students know what GIS stands for, it should be true regardless of the spelling—but some of the brightest students would miss it, due to the misspelled word. Item 2 is good for a true-false item because it exactly states a definition provided from the text materials and includes no distracting information—it is a "true" item. Item 3 is a typical, acceptable true-false item operating at the lowest level of RBT. Item 4 is also a good true-false item, but classifying its level of operation according to Bloom's taxonomy is not as simple because it depends upon how the information was taught. If cleaning up the Chesapeake Bay was a major topic of discussion in the text or lessons, then it operates at the lowest level of knowledge. However, if no specific mention was made of water pollution or tracking of pesticides and students were being presented a novel situation that they should figure out on the basis of similar topics, then the item tests at a higher level. It is possible to use true-false items to test some higher-level skills, but care must be used in doing so.

Multiple-Choice Items

Most evaluation textbooks recommend multiple-choice items as the most versatile and best form of item to use. That advice is correct when the items are developed by skilled authors, proofread carefully, edited, and then evaluated for validity and reliability (Ellsworth et al., 1990; Hoepfl, 1994). But few classroom teachers have either the time or the resources to develop their tests in this way. Therefore, as research by Haynie (1992, 1997a) recommended, technology teachers who choose to develop their own multiple-choice items must use care. Some important points to consider include:

- The stem (the question or statement portion of the item) must contain most of the information, and the alternatives (the answer choices) should be short to lessen the reading load on students. This is important because students forget what they have read when they must pore over lengthy alternatives. This penalizes slow readers, even if they know the subject matter well.
- Grammar, punctuation, and word choices must be accurate so that all of the alternatives correctly complete a sentence. This includes care in the use of articles, such as "a," "an," and similar words at the end of the stem. Students who are good readers will be able to eliminate some alternatives due to grammar problems even if they do not know the subject of study.
- All of the distractors (alternatives which are incorrect answers) must be plausible enough so that a student with incorrect or partial knowledge will view them as possible answers.
- Although some multiple-choice items ask students to choose the "best" answer from a field of alternatives in which several may be technically correct, in teacher-made tests it is advisable that only one of the

Conventional Classroom Assessment Tools: Item Selection, Development, and Evaluation

alternatives for each item be correct. The "best-answer" approach is an acceptable testing strategy only when the tests are developed with the care and field testing employed by skilled professional test developers. When classroom teachers attempt to implement this technique, the resulting items frequently are unfair, ambiguous, tricky, or simply not valid.

- There (optimally) should be one correct response and four distractor alternatives. If it is not possible to have at least four alternatives, the item is little better than a true-false item in terms of its capacity to identify those who really know the material.
- Style may vary, but consistency is important. If capital letters are used to identify alternatives, do so consistently and arrange items so they may be read easily. Items should not begin on one page and bleed over to another page—this includes the alternatives for the item. Readability issues such as these are especially important for computer and online tests because switching screens is not as seamless a process as turning a page.

Figure 2. Sample multiple-choice test items.

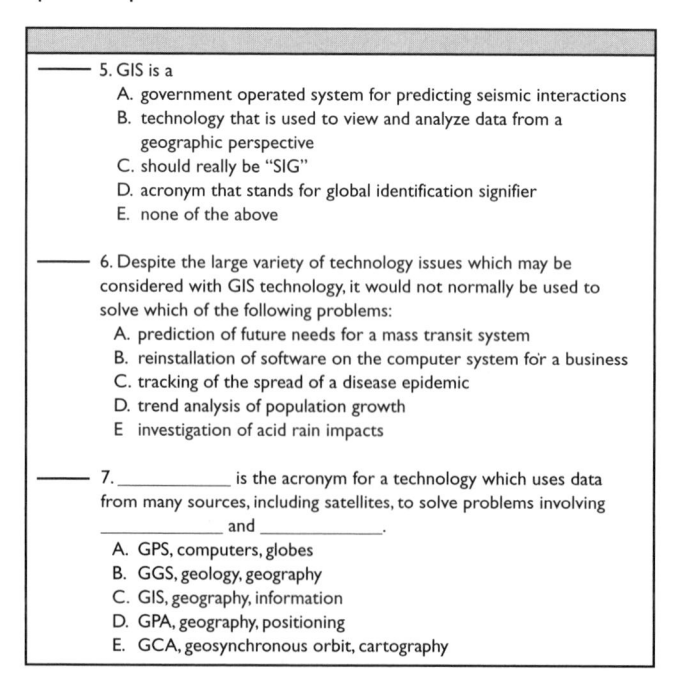

_____ 5. GIS is a
    A. government operated system for predicting seismic interactions
    B. technology that is used to view and analyze data from a geographic perspective
    C. should really be "SIG"
    D. acronym that stands for global identification signifier
    E. none of the above

_____ 6. Despite the large variety of technology issues which may be considered with GIS technology, it would not normally be used to solve which of the following problems:
    A. prediction of future needs for a mass transit system
    B. reinstallation of software on the computer system for a business
    C. tracking of the spread of a disease epidemic
    D. trend analysis of population growth
    E. investigation of acid rain impacts

_____ 7. _____ is the acronym for a technology which uses data from many sources, including satellites, to solve problems involving _____ and _____.
    A. GPS, computers, globes
    B. GGS, geology, geography
    C. GIS, geography, information
    D. GPA, geography, positioning
    E. GCA, geosynchronous orbit, cartography

Consider the examples of multiple-choice items shown in Figure 2. Question 5 demonstrates several common flaws. It was intended to operate at the "remember" level of the RBT cognitive dimension. The first problem is that the stem is shorter than the answer alternatives and does not provide a clear indication of the point of the question. Students will be so confused by the time they read all of the alternatives that they will need to re-read all of them in order to sort out the correct one, which is a great burden for slower readers. Grammatically, this stem can only go with alternatives A, B, or E. There is also no punctuation at the end of the stem; most frequently, a properly worded stem will end with either a colon or a question mark. The alternatives are not homogenous; they are like mixing apples and wrenches rather than apples and oranges. Is the basic question whether students know the acronym, or what GIS technology can do? What about alternative E? Many teachers make the mistake of throwing either "none of the above" or "all of the above" into any item for which they cannot conceive more than two or three alternatives. In those cases, students readily determine that the final choice is never actually the correct one. If those response alternatives are ever used, they should sometimes be the correct answers. This leads to two immediate results: First, the item automatically goes up in cognitive level. Second, it becomes more complex and difficult to develop. Students tend to dislike these items, and research by Haynie (1992, 1997a) found that items of this type developed by classroom teachers had more errors than any other forms. Worse still are items with an alternative such as "A and C but not B." Another pitfall for many teachers is unnecessarily stating an item in the negative, as shown in Question 6 (Figure 2).

In Question 6 it would be easy for a student to become confused and, while quickly reading to finish the test in the allotted time, fail to notice the word "not" in the stem. Since positively stated information is the norm, unless there is no other way to phrase the item, use positive statements exclusively. When you must resort to a negative statement in the stem, take two very important precautions: accentuate the word or phrase that renders the stem negative by underlining, using bold typeface or larger typeface, or even using all caps; and be very careful to avoid wording any of the alternatives negatively so that it would result in a double-negative when chosen. The above item would likely be acceptable if it read "it would **NOT NORMALLY** be used," but a little creative thinking by the test developer could provide a better item using only positive statements. A good quality of this item is that it does involve testing at the understanding or application level of RBT (depending upon how the information was taught).

Multiple problems are illustrated in Question 7 (Figure 2). Beginning the stem with a blank is poor because the stem should clearly reveal the main point of the question, and a blank first word is confusing. Multiple blanks in the same item also confuse and could trick students who know the material or provide easy correct guesses for students with partial knowledge. Alternative "E" does not fit the blanks because one part of it is two words. Actually, putting blanks in a multiple-choice item is rarely the best approach; item stems that end with each alternative completing the sentence are more often preferred. The alternatives in this example are weak, as well. If they all had "GIS" first, any of them potentially could be correct, so why have such a difficult and confusing item to determine if students know this one simple bit of jargon? Teachers should examine their first draft of each item carefully to ensure that items are not confusing for students. In short, keep multiple-choice items as simple in form as possible.

Matching Items

When a long series of very closely related multiple-choice items seems to evolve from the test blueprint, a matching item may be a better solution. This is especially true when similar stem-and-answer alternatives are being used repeatedly. Proper matching items consist of two lists in column form. Typically, the stems (questions) should be on the left and the responses (answers) should be on the right. The list of responses should include a few more choices than there are stems, to prevent guessing by elimination. Though some teachers break this rule, the response choices should not be used more than once each so that students may mark through them as they are used, to lessen their reading burden. As with multiple-choice items, the stems should be the longer statements, and responses should be very short to facilitate scan reading. Number the stems and label the responses with capital letters. Matching items are easy to score, students usually enjoy them, they can possibly test at upper levels of the cognitive domain, and they can help break the monotony of a lengthy test by providing stimulus variety. But there are many pitfalls to avoid. The first is long groupings: keep them in the 8- to 15-item range or they become a burden. Another mistake is to combine unlike information into matching items. This practice makes the items confusing for all students, it can provide easy opportunities for guessing by students with some knowledge, and it results in items that are less valid than other types of items would be. Consider the example shown in Figure 3.

Figure 3. Sample matching item.

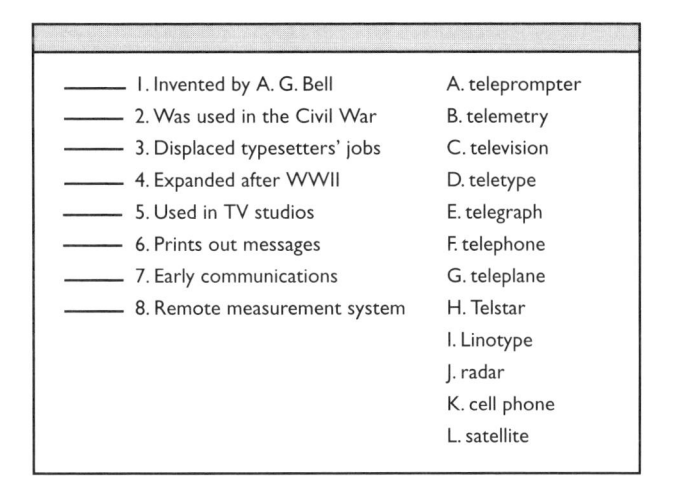

_____ 1. Invented by A. G. Bell	A. teleprompter
_____ 2. Was used in the Civil War	B. telemetry
_____ 3. Displaced typesetters' jobs	C. television
_____ 4. Expanded after WWII	D. teletype
_____ 5. Used in TV studios	E. telegraph
_____ 6. Prints out messages	F. telephone
_____ 7. Early communications	G. teleplane
_____ 8. Remote measurement system	H. Telstar
	I. Linotype
	J. radar
	K. cell phone
	L. satellite

The set of matching items in Figure 3 does have the sort of homogenous nature that enables use of matching items. To add in stems such as "cooks popcorn in 3 minutes," "forms a 3-D image," or "joins metal via cohesive bonding" simply to lengthen the set would reduce effectiveness and validity. Although these are reasonably acceptable, could some bright students be tricked into choosing answer "J" for item number 4? It is clear from the general thrust of the series that the intended answer is "C," but radar did expand greatly following World War II, particularly in its nonmilitary applications. The advice regarding matching items is to only use them when they truly are the best choice.

Essay Items

Essay or discussion items have many advantages and disadvantages from both the teacher's and the students' perspectives. They are generally easy for the teacher to develop but difficult to score. Their subjective nature compromises efforts to be fair in scoring. The length of student answers, often combined with poor handwriting, creates a large time burden for teachers.

Conventional Classroom Assessment Tools: Item Selection, Development, and Evaluation

From the students' perspective, they require a large burden of time to answer, but they allow students who are willing to devote enough effort to clearly evince all they know about the subject at hand. There are few cues to jog one's memory, so the successful answer reveals both remembering and understanding; often essays also test application to novel situations. Additionally, they reinforce the total school mission to teach writing. The freedom-of-response characteristic is one that wise students quickly learn to exploit when they do not fully know the correct answer because they can "explore" around the topic, groping for points by including related information that does not directly answer the question at hand. Students who write well and whose native language is English have an obvious advantage in answering these items, and oral responses may be the only fair alternative for students with reading/writing difficulties. It is also common for teachers scoring tests to develop an unintended bias toward answers they read first, which affects how they score the answers that follow. The first technique to increase the objectivity (and fairness) of essay items is to carefully develop an answer key as the item is conceived. Be sure to indicate to the students how many points the item carries. Each point should be clearly tied to a meaningful fact or concept in the model answer indicated in the key. Often teachers fail to consider this and simply assign the essay item(s) at the end of a test some numerical value to round the test out to an even 100 points, rather than carefully assigning the values based on what information a perfect answer would require. It should be acceptable for an essay item to have a value of seven points if it only requires six facts and one concept to create a perfect answer. When scoring the test, the teacher should mark all students' responses to item one, then shuffle the papers and mark all students' responses to item two. This process is repeated with subsequent essays, and the teacher should avoid looking at the names of students (though distinctive handwriting often foils this attempt at objectivity). Quickly scan each response, looking for the key words or phrases (or their appropriate synonyms). Write a number, beginning with "1," after each correct element you find. The last number that you write is the student's total score for that item. Some teachers prefer to mark correct elements with check marks and then count the marks and enter a number at the end of the response. Do not give credit for any response elements that were not on your answer key, regardless of how plausible they might seem, unless you are prepared to go back and review all students' papers to give them the same credit if they included this new fact.

Figure 4 shows an essay item from a technology education methods course on how to score essay items and a typical student response marked according to these guidelines.

Figure 4: Sample essay question, with student response marked by the teacher.

> 3. Briefly explain how to score an essay item. (5 points)
> Key for Item 3:
> Outline the key while writing the items
> Score item 1 for all, shuffle, then 2, etc.
> Avoid names
> Mark points in sequence
> Do not credit any "bull"
>
> Student's Response:
>
> 3. Make an outline of the answer you want when you write the test. **1** The outline should be brief and easy to use but no one but you will see it. Grade everyone's question 1 first and then do all 2's **2** without looking at names. **3** Don't give credit for extra garbage. **4** Check spelling & punctuation for each item & take points away for incomplete sentences.

As shown, the teacher found four correct elements in the student's response. This student included a novel element in his answer that may be both true and equally as important as other elements indicated on the teacher's model answer key. Since it was not on the key, it does not receive a point. However, the notion raised by this student is something for the teacher to consider when grading essays. One approach would be to deduct points for misspelled "technical" words that are clearly related to the topic at hand (with advance notification to students that they are responsible for learning to spell them) and to simply mark other grammar and spelling errors without deducting points to help students learn to recognize problems in their writing. In nontest writing assignments such as term papers, PowerPoint® presentations, and lab reports, such errors should be marked with a penalty to reinforce writing, but not on the test.

One more consideration is the choice of whether to deduct points for incorrect statements that are made in addition to the correct statements required for an answer. If the preceding guidelines for scoring essay items are followed precisely, incorrect statements would simply be ignored when scoring. Many teachers would argue that a student who made four correct statements and then added an additional incorrect statement (possibly searching for more points) should receive a grade of three rather than four, to penalize for the incorrect statement. That practice, if it is known to the students, does help prevent students from artificially lengthening their answers. It also may help students learn a more efficient writing style that could benefit them in other contexts. In any case, teachers who intend to deduct points for wrong statements should clearly inform students before they take the test.

Short-Answer Items

Short-answer items have three fundamental forms: brief questions similar to essay items but only requiring a few words or one sentence to answer; statements with one or more blanks for students to fill in; and items that appear as multiple-choice item stems without any response alternatives, so that students must answer from recall. In all cases, the stem must be clear enough to elicit only the desired response. Generally, the guidelines for developing good short-answer items vary, depending upon which form is used, but the factors explored in the sections above should help teachers improve their short-answer items. The greatest danger is an ambiguous stem that may be answered correctly in multiple ways or that does not clearly lead students to the desired answer. Since teacher-developed multiple-choice items are often spoiled by confusing and contradictory answer alternatives, teachers who have difficulty developing effective multiple-choice items may consider short-answer items as a good alternative. Avoid confounding items with multiple blanks in the same item. Figure 5 shows examples derived from earlier items.

Figure 5. Sample short-answer test items.

> 8. What do the letters in the acronym "GIS" represent?
>
> 9. Identify one application of GIS technology.
>
> 10. Identify a technology system that could help a government in its efforts to keep pesticides out of a body of water such as the Chesapeake Bay.

Item 8 is a clear replacement for previous example true-false item number 1 or number 3 (depending upon the teacher's intent). Its only acceptable answer is "Geographic Information System." Had the item simply been stated "What does GIS stand for?" then this would be an acceptable answer along with several others such as "a system that uses geographic information to solve problems." Item 9 replaces true-false item 4 or multiple-choice item 6. There would be many acceptable answers. If one particular answer is desired, the stem must be expanded to limit possible answers. In answering item 10, both the acronym and the full name would be acceptable answers, unless the stem is altered. The essential consideration is to ensure that the stem elicits only the exact desired answer, unless a range of answers is acceptable.

Problems

Many of the guidelines for improvement of essay items also apply to story problems or word problems. The essential consideration is to ensure that students understand the intent of the item. The number of points credited for the various steps toward solution and the number given for the correct answer must also be known to the students. Direct students to show their work and indicate whether the answer should be stated in a complete sentence, circled, or otherwise presented. Whenever a common hypothetical situation or data set is to be used for several problems, make it clear to students that this is being done and cluster all of the related problems together on the test. Make it a habit to deduct points when students fail to show their work, for two reasons: The algorithm (process toward solution) is often as important as the final answer, and awarding full credit for the final answer alone makes it more difficult to detect cheating by students. The quality of problems on teacher-made tests may be improved by asking a mathematics teacher to review and discuss them prior to final revisions.

ADMINISTERING TESTS

Even when a well-developed test is available, its effectiveness for evaluating student learning may be hindered by common oversights or confusion in the administration procedures or the environment. Preparing for testing includes preparing the instructions and the testing environment, as well as encouraging students to be prepared for the test.

Maintaining a Good Testing Atmosphere

Some students fear tests and view the test environment as a tense and threatening situation. Unfortunately, some teachers exacerbate this unhealthy climate by making threatening statements or by adopting an uncommonly authoritative demeanor on test days. Tests should not be viewed as major events—they should simply be one aspect of the educational process. Yes, the room must be quiet to allow students to concentrate, to reduce confusion, and to limit cheating, but it should not seem hostile.

Do not begin the test until everyone is seated and has the needed materials (paper, pencil or pen, cover sheets, references if allowed, and others as needed). Ensure that all students have adequate space, can easily shield their paper from view of others, are comfortably situated, and that the room is quiet, well lighted, and well ventilated. It is generally advisable to make any needed announcements and to give general directions (such as how to submit papers) before distributing the test copies because many students will focus only on the test itself once they have it in hand. If you have discovered errors in the test copies, note them on the chalkboard, screen, or whiteboard and call attention to them before the distribution of tests. Remind students that the test is important, but do not make statements to induce fear. Make a last call for full attention and continued quiet prior to distributing the copies. Have some sort of quiet assignment outlined for students who finish early. After students receive copies of the test, encourage them to look the test over briefly and pace their work. Ask if there are any general questions before beginning.

Once the test begins, avoid interruptions. When students ask questions about the test or individual items, determine before answering if the question is valid and its answer does not reduce the validity of the item. If you decide to answer the question, do so for the whole class by calling their attention to the item, repeat the student's question, and then answer it for all to hear. Add this to the errata notations on the board and mention the item to the next class prior to the test. Have clear instructions on the test concerning how to answer each type of item, the point values of each section and the whole test, where to write, and how the students should identify their papers (names or numbers and where to place them). Also, have written instructions about where to submit papers, whether to clip or staple, and other factors, so that you will not have to speak often during the test (these may be on the board with the errata). Ensure that there is adequate time for the test. Never rush students to finish in a certain allotted time, unless it is a specific goal of the test, in which time is a valid measure of job proficiency (e.g., a timed test in typing).

Maximizing Learning Benefits of the Test

Several research studies by Haynie (1990b, 1991, 1994, 1995, 1997b, 2003a, 2003b, 2004) have found that, in addition to providing information for evaluation, tests aid the learning process. Students learn while they prepare for a test. In addition to this motivational aspect of the testing process, the actual act of taking a test has been shown to lead to greater learning gains. In these studies, students who were informed of an upcoming test performed better than those who were told they would not be tested, and students who then actually did take a test learned more than those who did not receive the anticipated test. These results were confirmed by student performance three weeks later on an unannounced test for delayed retention. Other means to increase the learning value of a test include what is done before and after the test. Reviews of information before the test help students retain longer. So does a review of the right answers, with students examining their own scored papers, after the test. Having students research to find corrections for their own errors is another common way to turn the test into a learning aid. Review sheets and preparation or study questions given in conjunction with tests have also been shown to aid in retention learning (Nungester and Duchastel, 1982). One study even found that the best form of test to use to maximize learning retention is a take-home test. However, this same study showed that students only learn the precise facts and information actually represented on a take-home test, while they study more broadly and learn related information better when tested in class (Haynie, 2003b). Tests require far too much teacher and student time not to maximize these beneficial learning effects.

Encouraging Honesty

One troubling concern for teachers is cheating. The professional literature is full of examples of dishonesty among students, and virtually every teacher has had to confront a student about cheating. A wise old saying is that "a lock is only to keep an honest man honest." The implication is clear: a committed thief will simply break the lock, but the lock will limit the strength of the temptation to steal for most people. The same is true in the classroom. The teacher should arrange seating to make cheating more difficult, carefully monitor students to watch for note-passing and whispering, encourage use of cover sheets, challenge students when their papers have exactly the same wording in response to essay answers, and keep the room quiet and orderly in general. In the final analysis, the truly dishonest student may still find a way to cheat. The teacher should use a positive approach as much as possible by encouraging honesty rather than threatening to punish if cheating is discovered. Help all students remember to do their own part to keep papers covered (like putting a lock on a door), and provide a safe means for students to report peers whom they fear have cheated. Students will recognize diligence in establishing order and in monitoring, and that will go a long way toward ensuring honesty.

ANALYZING CLASSROOM TESTS FOR IMPROVEMENT

Standardized tests undergo extensive review and editing during their development and then are subjected to careful scrutiny via statistical analysis of data collected during pilot testing. Once final revisions are completed and the test is actually published and placed in service, data collection continues to develop pools of information for interpreting the test results normatively as the characteristics of the populations being tested evolve. Although it would be impractical or impossible for classroom teachers to conduct the kind of review and analysis done to standardized tests, there are some very practical ways they can examine and improve the tests they develop. With the ease of revision provided by modern word-processing software, there is little excuse for continuing to use a test that is found to be flawed.

Validity and Reliability

There are several types of validity, but the type most commonly referenced is content validity, which measures the extent to which a test actually tests what it is intended to. In layman's terms, we could use "truthfulness" to refer to content validity. As an example, suppose a communications technology teacher administered a test about CAD applications, but the test focused on terms and features found in a CAD program not used in the class. It might happen that there was one student who had used the alternative CAD program and who knew enough about it to score reasonably well on the test, while all others, even those with good knowledge of their classroom version of CAD, scored poorly. Do those results truthfully reflect the students' knowledge of CAD software? For a test to be valid, each item on the test must also be valid. That means each item must match well with what is intended by the course objectives and the test blueprint.

Reliability is a different, but related, issue. To say that a test is reliable essentially says that the test will consistently render the same results if used again and again in a given situation. Basically, if the same test is used in several class sections or over a period of semesters, do the results generally seem stable, or is there wide variation of scores with each administration? One might ask how reliability and validity relate. A test must be reliable if it is to be valid—a test simply cannot be telling the truth if its results are different each time it is used. Being reliable does not automatically ensure that a test is valid. Recall the example of the test focusing on an alternative CAD program. The test was probably not truthfully assessing students' knowledge of CAD, but it would likely be highly reliable because the scores would not vary much if we gave the same test again the next day or when we used it in the third-period as well as fifth-period class. So, it would be reliable (consistent), but still not be valid (truthful).

How can technology education teachers improve the validity and reliability of the tests they develop? If we concentrate on the validity of each item on the test, overall validity will likely be enhanced, and reliability should follow as a result. So, the advice here is to strive to ensure that every item is valid. This begins by careful attention to the development of the test blueprint so that it properly reflects the content and objectives being tested. This will ensure the overall validity of the test itself, and, if care is taken to make each item within the test valid, a good test is assured. Once the test is actually used, however, some real data can be analyzed to improve the test when it is revised. Item analysis is the best technique for classroom teachers to use in revising their tests.

Item Analysis

A simplified form of item analysis is helpful to the technology teacher who wishes to analyze tests for improvement. Divide the papers into two groups according to their scores on the total test: upper scores and lower scores. We will assume that multiple-choice items are being analyzed in this example, but the procedures may be easily altered for other types of items. The main point is to determine if each item is being answered correctly by the better-prepared students and missed more often by the weaker students. For multiple-choice items it is also simple to determine which distractor alternatives are operating properly and which ones are weak. Hypothetical data is recorded for analysis of example items 5 and 6 (Figure 2). A class size of 22 students is assumed. Since 10 is an easy number to work with, the two scores in the middle are ignored for convenience and the 10 highest-scoring and ten lowest-scoring students' papers comprise the upper and lower groups. Table 3 shows a portion of the spreadsheet with the information recorded. Bold typeface and an asterisk indicate the keyed correct answer for each item.

Table 3. Raw data recorded in a spreadsheet for item analysis.

Item Number	Score Groups	A	B	C	D	E	Omit	Difficulty %	Discrimination Power
5	Upper	0	10	0	0	0			
	Lower	1	3	2	3	1			
			*						
6	Upper	4	2	2	1	1			
	Lower	5	1	1	2	1			
			*						

Once the data are recorded, the formulas for calculating difficulty and discriminating power are simple:

Difficulty = $\frac{\text{Correct Answers}}{\text{Total Attempts}}$ (Record %)

Inserting the data for Item 5 we find:

Difficulty = 13/20
Difficulty = .65   Record 65%

Examining the formula helps the reader understand what the resulting value indicates. An item that is so difficult that no students choose the correct answer will have a difficulty rating of 0% (extremely difficult). An item correctly answered by only one student is rated 5%, and one that all students answer correctly receives a rating of 100%. The 65% rating of item 5 shows a moderately difficult item. The test should have items with a range of difficulty ratings from about 25% to 75%, with many items hovering near the 50% level. A test with many highly difficult items is too frustrating for students, while tests with too many easy items may be of little value for evaluation purposes. Table 4 shows the same table of data, with the calculated values for both difficulty and discriminating power inserted.

Conventional Classroom Assessment Tools: Item Selection, Development, and Evaluation

Once difficulty is found, solving for discriminating power is simple, too. Here is the formula:

$$\text{Discriminating Power} = \frac{\text{Correct Upper Group} - \text{Correct Lower Group}}{\text{Total Attempts}/2}$$

Inserting data for Item 5, we find:

Discriminating Power = (10-3)/(20/2)
Discriminating Power = 7/10
Discriminating Power = .70

Careful analysis of this formula shows that an item that is marked correctly by only the students in the upper group will have a discriminating power of 1.00. Similarly, an item that every student answers correctly will be rated 0.0, and items with discriminating power ratings that are negative numbers are those that the brightest students answer incorrectly, but the lower group students answer correctly. These items should be deleted or heavily revised before they are used again. Discriminating power ratings near .50 are generally good for most items on teacher-developed classroom tests, and a range from about .25 to .75 is common. Example item 5 discriminates effectively, while item 6 is so difficult (15%) that it probably frustrates all students in the class and its discriminating power (.10) is poor as well. It appears that there is some flaw in distractor A of item 6, attracting the best prepared students away from the correct answer.

Table 4: Difficulty and discriminating power recorded.

Item Number	Score Groups	A	B	C	D	E	Omit	Difficulty %	Discrimination Power
5	Upper	0	10	0	0	0			
	Lower	1	3	2	3	1			
			*					65	.70
6	Upper	4	2	2	1	1			
	Lower	5	1	1	2	1			
		*						15	.10

If the same items are used a second time (with or without revision), the process of item analysis may be repeated and a history developed for each item. When used in conjunction with careful assessment of content validity, perhaps by having colleagues review the items, along with the teacher's own careful proofreading and editing, these simple procedures are adequate for improving teacher-developed classroom tests.

Research Findings on Teacher-Made Tests

Haynie (1997a) found that technology teachers differed greatly in their ability to develop effective test items. These findings replicated, in most regards, those of his 1992 study. Problem areas included grammar, punctuation, mismatch of items to objectives (both in content and level), validity, clarity, and poor distractor alternatives used in multiple-choice items. Haynie also found that experienced teachers (those with more than eight years of service) outperformed novices, and both undergraduate and graduate courses in tests and measurements were helpful for improving teachers' test-development skills. However, not all undergraduate programs include them. Experienced teachers with master's degrees and at least one course in tests and measurements developed the best items. The most notable change in the quality of items developed in the two studies was that spelling errors, which were very common in 1992, were nearly nonexistent in 1997. Haynie attributed this finding to the fact that the teachers in the later study all had used computers with spell-checking software. There was, however, a new problem noted in the second study. Although the finding was not statistically significant, there were more frequent occurrences of the wrong word being used when two or more words sound alike (e.g., "there" was used when the proper choice would be "their" or even "they're"; or "an" and "and" were confused; or "transmit" replaced "transit"). In all of those cases, the error was not detected by spell-checking software, and it appeared that the item author had proofread less carefully than he or she would have in the era prior to the availability of such software. So, the primary caution in development of test items is certainly to proofread them carefully, rather than relying on the computer to find and repair all problems.

Testing Using Computers

In addition to word-processing software with spell-check capabilities, there are many other computer applications now available that assist in the development of tests. As with any computer function, the old motto "garbage in=garbage out" applies. The computer will not improve weak or invalid items that are entered. Teachers must periodically check item banks to ensure that the items continue to operate as desired and continually update the item banks as appropriate. The more powerful software packages permit test generation, selection by difficulty or grade level, coding by standards, and many other useful features. An item bank needs to be large enough so that multiple forms of tests on the same content are available, and items may be used to generate other resources, as well as tests (e.g., review sets, pretests, study packages, etc.). When whole districts or states use a large item bank, it is possible to perform sophisticated analysis, and the resulting tests may be normed, as is done with standardized tests. This makes the information gathered by the tests more useful, but does possibly detract from the unique nature of individual teacher-made tests. The ability to be customized for an individual classroom or setting is lost as the items and resulting tests become standardized (and thus, homogenized).

When students actually take their tests on computers, instant results may be given, which could help students learn more and retain the information learned. Computerized and online testing permits use of many features not possible with paper-and-pencil tests, such as multimedia based items, 3-D animations, color graphics in illustrations, rotation of 3-D images, video clips, sound, and more. These bring a reality to the testing situation that would have been inconceivable in the past. Vendors are already beginning to market testing software, tests, and other media with some of these features. Two precautionary statements are timely concerning increased use of these computer-based testing techniques and advanced features. First, the prudent teacher will take care to ensure that testing software, applications, and features selected do not outpace the availability and capabilities of the school's computer systems and networks. Second, since many of these features require the computers to be networked, increased care must be taken to ensure that students are not able to cheat by accessing information directly from the Internet during the test or by temporarily installing their own media on the machine that they use.

LIMITATIONS AND PITFALLS OF TRADITIONAL CLASSROOM TESTS

Although a good case has been made here for the importance of classroom tests in modern technology education, limitations must be recognized. Tests are not capable of evaluating many important skills in a modern technology class. Written tests are generally not effective for evaluation of projects, learning products constructed by students, reports, presentations, role-plays, and daily work habits of students. Teachers' observations are a far better assessment of students' adherence to proper laboratory procedures and safety rules than any type of items appearing on written tests. Tests do not help teachers understand students' motivations, preferences, or dispositions. Rubrics, which are natural embellishments of the project evaluation sheets used in the industrial arts shops of yesteryear, yield far more meaningful evaluations of student work in many laboratory activities than do tests. The goal here is not to encourage technology teachers to use written tests in all situations, but to help them improve the tests they develop when a test is the best choice for evaluation of the skill or knowledge under consideration.

In these days of increased use of high-stakes, end-of-course testing in many subjects, there is fear within the education community that teachers will "teach to the test" in an effort to improve their students' scores. These fears are well-founded, and the damage such practices could do to the total education program are significant. Teaching specifically and exclusively what is on the end-of-course test may lead to sterile instruction that is devoid of character and innovation; to tedious assignments, such as drill and practice; and to inhibition of creativity for both students and teachers. One of the benefits of teacher-made tests is that they can be tailored specifically to what is being taught in a particular classroom or setting (Mehrens and Lehmann, 1987). Technology teachers are encouraged to seek creative approaches, interesting projects, and a variety of ways to evaluate student performance.

CONCLUSION

This chapter focused on tests developed by technology teachers. Precautionary statements were presented concerning the limits of the applicability of tests, teachers' preparedness in test development, and applications of various types of test items.

Development of an effective test begins when a test blueprint based on the learning objectives is developed. This ensures balance and also helps to align the test items with both the information under study and the types and levels of objectives. Careful proofreading is essential, and having a colleague review the test items and comment on their content validity is also helpful. Guidelines for improving the quality of various types of test items were presented. The testing environment should be controlled and businesslike, but not tense. The teacher who wishes to improve a test can easily conduct a simple analysis of each item for difficulty and discriminating power. There are limits to what should be assessed by tests versus by other means.

Since many technology teachers do not take tests-and-measurements courses as part of their undergraduate degree programs, the burden for helping them understand how to develop effective classroom tests falls on the faculty teaching their methodology courses (Haynie, 1992, 1997a). Even when undergraduate students do take a tests-and-measurements course, there is no guarantee that topics concerning how to develop and analyze effective teacher-made classroom tests will be adequately covered. They may instead be generalized courses that stress a lot of information concerning standardized tests and how to interpret their results. Even when those broadly defined courses do include information on development of teacher-made tests, they typically do not make any specific mention of the unique situations found in technology laboratories. Tests-and-measurements courses are suggested as required components of graduate programs in technology education. The master teachers and future leaders in our profession should have increased background to help them understand development, norming, analysis, and interpretation of standardized tests and to also help them gain a clear understanding of how to apply portions of that information to improving teacher-made tests. Wherever such courses are not available or other curricular demands are considered more important, the technology education faculty must ensure that graduate students leaving programs are well prepared to develop and analyze effective classroom tests.

DISCUSSION QUESTIONS

1. Classroom tests are only one means of evaluation in technology laboratories. List the major topics, skills, and knowledge studied in your classroom, and cite the appropriate methods for evaluation of each. Which ones are best evaluated with teacher-made tests?
2. Make a case for utilizing teacher-made classroom tests as learning activities. What are some ways to maximize their effectiveness as aids to learning?
3. What are the most common problems found in teacher-made classroom tests? How might you eliminate these problems from the tests that you develop?
4. Analyze this statement and explain the meanings of the two terms (*reliability* and *valid*): "Reliability is a necessary but insufficient element of a valid test." Is the statement true or false? Why?
5. Analyze the most recent teacher-developed or vendor-supplied test you used in your technology classes according to the guidelines provided in this chapter. What common problems did you find, and how can they be remedied?
6. Despite the many problems found in some teacher-made tests, the author of this chapter encourages their use. Make a compelling argument advocating continued use of teacher-made classroom tests and also cite potential drawbacks. Does the increased use of required end-of-course tests in many states alter the argument? If so, how?

REFERENCES

Anderson, L., D. Krathwohl, P. Airasian, K. Cruikshank, R. Mayer, P. Pintrich, J. Raths, and M. Wittrock, eds. (2001). *A taxonomy for learning, teaching, and assessing: A revision of Bloom's taxonomy of educational objectives.* New York: Addison Wesley Longman Inc.

Bertrand, A., and J. P. Cebula (1980). *Tests, measurement, and evaluation.* Reading, MA: Addison-Wesley.

Bloom, B. S. (1956). *Taxonomy of educational objectives: Handbook I, cognitive domain.* New York: David McKay Co.

Burdin, J. L. (1982). *Teacher certification. In Encyclopedia of education research,* edited by H. E. Mitzel., 5th ed. New York: Free Press.

Carter, K. (1984). Do teachers understand the principles for writing tests? *Journal of Teacher Education* 35(6): 57-60.

Ellsworth, R. A., P. Dunnell, and O. K. Duell (1990). Multiple choice test items: What are textbook authors telling teachers? *Journal of Educational Research* 83(5): 289–293.

Fleming, M., and B. Chambers (1983). Teacher-made tests: Windows on the classroom. In *Testing in the schools: New directions for testing and measurement,* edited by W. E. Hathaway (no. 19, 29–38). San Francisco: Jossey-Bass.

Gronlund, N. E. (1981). *Measurement and evaluation in teaching.* New York: Macmillan.

Gullickson, A. R., and M. C. Ellwein (1985). Post hoc analysis of teacher-made tests: The goodness-of-fit between prescription and practice. *Educational Measurement: Issues and Practice* 4(1): 15–18.

Haynie, W. J. (1983). *Student evaluation: The teacher's most difficult job.* Monograph series of the Virginia Industrial Arts Teacher Education Council, Monograph Number 11.

——— (1990a). Anticipation of tests and open space laboratories as learning variables in technology education. *Journal of the North Carolina Council of Technology Teacher Education* 1(1): 2–19.

——— (1990b). Effects of tests and anticipation of tests on learning via videotaped materials. *Journal of Industrial Teacher Education* 27(4): 18–30.

———(1991). Effects of take-home and in-class tests on delayed retention learning acquired via individualized, self-paced instructional texts. *Journal of Industrial Teacher Education* 29(1): 1–12.

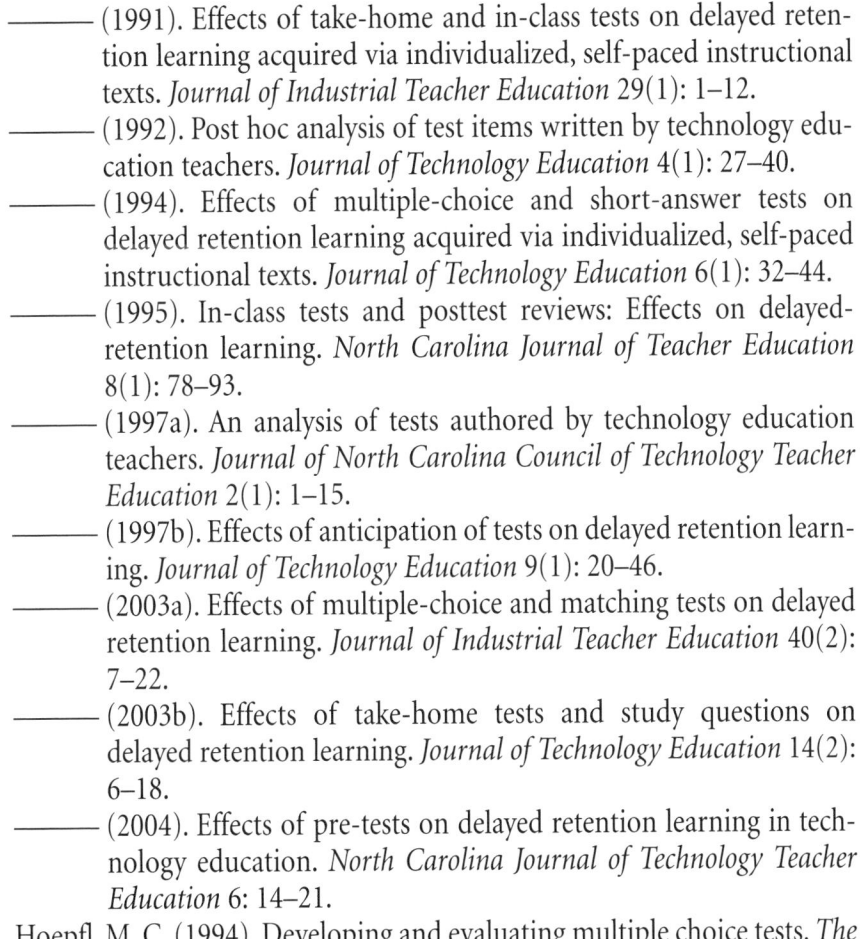

———(1992). Post hoc analysis of test items written by technology education teachers. *Journal of Technology Education* 4(1): 27–40.

———(1994). Effects of multiple-choice and short-answer tests on delayed retention learning acquired via individualized, self-paced instructional texts. *Journal of Technology Education* 6(1): 32–44.

———(1995). In-class tests and posttest reviews: Effects on delayed-retention learning. *North Carolina Journal of Teacher Education* 8(1): 78–93.

———(1997a). An analysis of tests authored by technology education teachers. *Journal of North Carolina Council of Technology Teacher Education* 2(1): 1–15.

———(1997b). Effects of anticipation of tests on delayed retention learning. *Journal of Technology Education* 9(1): 20–46.

———(2003a). Effects of multiple-choice and matching tests on delayed retention learning. *Journal of Industrial Teacher Education* 40(2): 7–22.

———(2003b). Effects of take-home tests and study questions on delayed retention learning. *Journal of Technology Education* 14(2): 6–18.

———(2004). Effects of pre-tests on delayed retention learning in technology education. *North Carolina Journal of Technology Teacher Education* 6: 14–21.

Hoepfl, M. C. (1994). Developing and evaluating multiple choice tests. *The Technology Teacher* 53(7): 25–26.

Herman, J., and D. W. Dorr-Bremme (1982). Assessing students: Teachers' routine practices and reasoning. Paper presented at the annual meeting of the American Educational Research Association, New York.

International Technology Education Association (2000). *Standards for technological literacy: Content for the study of technology.* Reston, VA: International Technology Education Association.

Mehrens, W. A. (1987). Educational tests: Blessing or curse? (n.p.)

Mehrens, W. A., and I. J. Lehmann (1987). Using teacher-made measurement devices. *NASSP Bulletin* 71(496): 36–44.

Newman, D. C., and W. M. Stallings (March 1982). Teacher competency in classroom testing, measurement preparation, and classroom testing practices. Paper presented at the annual meeting of the National Council on Measurement in Education (Mehrens and Lehmann, 1987).

Nitko, A. J. (1989). Designing tests that are integrated with instruction. In *Educational measurement*, edited by R. L. Linn. 3rd ed., 447–474. New York: Macmillan.

Nungester, R. J., and P. C. Duchastel (1982). Testing versus review: Effects on retention. *Journal of Educational Psychology* 74(1): 18–22.

Stiggins, R. J., and N. J. Bridgeford (1985). The ecology of classroom assessment. *Journal of Educational Measurement* 22(4): 271–286.

Stiggins, R. J., N. F. Conklin, and N. J. Bridgeford (1986). Classroom assessment: A key to effective education. *Educational Measurement: Issues and Practice* 5(2): 5–17.

Alternative Classroom Assessment Tools and Scoring Mechanisms

Chapter 4

Marie Hoepfl
Appalachian State University

INTRODUCTION

The term *alternative* is typically applied to any assessment other than the so-called traditional or paper-and-pencil test. Traditional assessments generally employ multiple-choice or some other type of closed-response question. Because these items require selection of a single, correct response, they are often perceived to be more objective than assessments with multiple, open-ended solutions. Alternative assessments are often performance based, and may or may not be considered "authentic." Thus, although it is not uncommon to see the three terms—authentic, performance, and alternative—used interchangeably (Walberg, Haertel, and Gerlach-Downie, 1994, p. 7), the more inclusive term *alternative* will be used predominately within this chapter.

In today's educational climate, views about assessment tend to be polarized. Assessment is seen as a "potent tool for change" in the schools and assessment that relies on traditional tests is an attractive option to policy makers because they are relatively easy and inexpensive to create and administer (National Research Council [NRC], 2002, p. 60). However, to many educators and parents, these standardized, often high-stakes tests lack instructional sensitivity, and their use can lead to negative changes in the teaching and learning process. "Teaching to a bad test and spending months on drill and skill may boost scores but surely ends up turning off students" (Lederman and Burnstein, 2006, p. 430). Some educators believe that open-ended alternative assessment tasks are more challenging and reflective of the desired learning outcomes and, as a result, that they give truer pictures of student understanding (Shepard, 1997). In reality, a balanced and comprehensive assessment plan will likely include both traditional and alternative measures of student capability.

High-quality assessment plans will emphasize the improvement of teaching and learning; will align with established criteria or standards, as well as with the local enacted curriculum (i.e., what is actually taught); and will include multiple measures of performance, including both traditional and alternative measures (International Technology Education Association [ITEA], 2004; Jones, 2004; NRC, 2001). The latter point is critical because no single assessment tool can provide all of the information needed about student cognition and performance, particularly in fields such as technology education where process skills are a key part of the learning objectives. Although paper-and-pencil tests can be used to capture cognitive understanding, such tests are unable to capture the dimensions of knowing related to the "doing" of technical tasks. The goal of alternative assessment is to wed paper-and-pencil content assessments to the richer performance data found in classrooms and to effectively round out the picture gained about teaching and learning in classrooms.

The goals of this chapter are threefold. It is designed to
- expand the reader's overall understanding of alternative assessment strategies and tools;
- give readers a better understanding of how to create alternative assessments and scoring mechanisms that elucidate and align with the content standards; and
- show readers ways that alternative assessments might be included in an overall assessment plan.

ALTERNATIVE ASSESSMENT TOOLS AND TECHNIQUES

Scott (2000) provides an excellent overview of alternative assessment tools for the technology classroom. He notes, "As one reviews the list of tools, it will become immediately obvious that there is scant distinction to be made between performance activities and assessment techniques" (p. 38). Scott divides his examples of assessment tools and activities into three categories: graphic organizers and concept mapping, performance products, and live performances (p. 39). Each category contains a wide sampling of activities that could be used to demonstrate conceptual understanding or process skills. What must be understood is that each of these sample activities, if used for assessment purposes, requires a set of criteria or standards against which the performance will be judged using some type of scoring mechanism. Such scoring mechanisms form the heart of alternative assessments, because they define the desired characteristics or attributes of performance that will be examined. For this reason, a great deal of attention will be paid throughout the rest of Chapter 4 to the development of scoring mechanisms.

Another helpful resource that provides an overview of alternative assessment tools and scoring mechanisms is the *Technology Education Performance Based Education Implementation Handbook* (Missouri Department of Elementary and Secondary Education, 1994). This handbook includes examples of a variety of types of scoring mechanisms, from checklists to holistic rubrics, which could be useful in the technology education classroom. Similarly, the ITEA document *Measuring Progress: Assessing Students for Technological Literacy* (2004) provides a sampling of assessment tasks, along with a description and discussion of the advantages and limitations of each. Readers will find all three of these documents useful additions to their professional bookshelf.

Performance tasks that might be used in the technology education classroom are virtually limitless, and could range from student presentations to the creation of a diagram that describes the function of a robot arm, or to assembling a mock-up that illustrates a design idea. The selection of any performance task that will form the basis of an alternative assessment should, however, be determined by one simple question: To what extent will the activity provide acceptable evidence that the desired results have been achieved? Answering this question requires having a good understanding of the curriculum standards and of what is most significant within those standards. It is only then that the actual activities that best match the curriculum goals, and that will elicit the acceptable evidence regarding goal attainment, should be selected (Wiggins and McTighe, 2001; Brown, 2004).

SCORING MECHANISMS

Alternative assessments generally contain elements that cannot be scored or assessed in traditional ways. That is, they generally involve the completion of a task that is multifaceted, requiring that multiple characteristics be examined. Usually there is not a single "right" answer; therefore correct responses can take differing forms. Levels of quality in completing the task can vary, but still represent satisfactory work. Therefore, some type of scoring mechanism must be created that captures the critical components of the assessment task, as well as its attributes. The term *rubric* is generally applied to such a scoring mechanism. There are, however, a number of variations for this type of tool.

Quinlan (2006) distinguishes between evaluation checklists, performance lists, and rubrics. She further identifies two types of rubrics—holistic and analytic. Checklists identify the components or steps that must be included in a performance product. These are typically binary lists that indicate whether a component is present or not present, and do not usually involve judgments of quality. Ratings on a checklist can be expanded somewhat by using a check (the component or step is present), a plus sign (the component is consistently present), or a zero (the component is not present) to indicate the degree to which a desired component is exhibited. Quinlan notes that, in spite of their limited capacity as scoring mechanisms, checklists can provide a foundation for more detailed performance lists and rubrics.

In situations where simple checklists are not sufficient, an enhanced scoring mechanism is needed. A performance list usually has a list of components or attributes, with a possible range of scores for each item in which the range of points is defined by some sort of scoring key (e.g., 4 points=excellent, 3 points=good). Their drawback is that they do not define, for students, the precise criteria used by the teacher in determining the numeric score in each category. To clarify the attributes of expected performance for students, and to be able to justify the numerical grade assigned, it is necessary to create a scoring rubric that details the exact criteria on which a score will be based (Quinlan, 2006).

Holistic rubrics are used to evaluate an entire project or performance and yield one overall score. Their advantage is that they can be easier to create and to score than other types of rubrics. However, they provide few details about the relative quality of various aspects of a performance, so their usefulness in providing feedback to students is somewhat limited. Analytic rubrics, on the other hand, break down the features of a project or performance and examine a range of characteristics for each feature. The number of components included can vary, depending on the complexity of the task. Analytic rubrics allow for more detailed analysis of performance, providing more specific feedback to students and teachers. Also, some components can be weighted more heavily than others, reflecting relative importance (Quinlan, 2006).

Detailed scoring rubrics provide many benefits compared with other scoring tools, and these add to their utility from both a formative and summative perspective. Aside from their great value in helping make what could be a fairly subjective analysis more objective and defensible, rubrics have other significant benefits. Probably most important is their capacity to inform. By creating a rubric, the teacher must break down a performance into its component parts and in the process make decisions about which are the most critical components. The teacher must also articulate the characteristics of each component and define what the range of performance might look like. When this information is shared with students, as it always should be at the start of an activity to which the scoring rubric will be applied, the rubric plays a formative role up front. Additionally, it enables the students to become engaged in self-assessment, because they will be able to perform self-checks as they progress through the activity prior to its completion (Quinlan, 2006).

Establishing Performance Outcomes and Scoring Criteria

The starting point for any alternative assessment scoring tool, just as with traditional assessments, is the learning objectives. These define what the critical components of a unit or lesson are. A helpful step in articulating the learning objectives is to develop an assessment framework. This framework will be based on the content standards and will graphically display the range and type of competencies desired. From this, it's possible to more systematically determine the nature of the assessment task or tasks to be used.

The work done by Lorin Anderson and his colleagues on the Revised Bloom's Taxonomy (RBT) provides one strategy for developing an assessment framework that can be used to better align assessments with objectives. The RBT table is a matrix consisting of four knowledge dimensions along one axis and six cognitive process dimensions along the other axis. Following the guidelines described by Anderson et al. (2001), teachers can identify the major objectives within a unit of study and place them in the corresponding matrix cells on the taxonomy table. Assessment tasks can similarly be placed in their corresponding matrix cells. Assessments and objectives will be aligned if they share the same cells on the matrix and if they place similar amounts of emphasis on each element. Although it is beyond the scope of this chapter to provide a detailed description of RBT, readers are encouraged to learn more about this approach and to apply it to their own instructional planning efforts.

Part of the challenge in aligning assessments with objectives is dealing with what Anderson et al. (2001) call the "level of specificity problem" (p. 105). That is, to what level of detail must we go when unpacking learning objectives for a particular unit or lesson?

> A useful test of the specificity of an objective is to ask whether, after having read it, you can visualize the performance of a student who has achieved it. "What would a student have to do to demonstrate that he or she has learned what I intended him or her to learn?" If you envision a variety of different performances, you probably ought to ask, "What performance is the most representative of the achievement of this objective?" Discerning this central performance narrows broad objectives down to the more specific ones. (Anderson et al., 2001, p. 105)

By focusing on the alignment between objectives, assessment, and instruction, we are better prepared to create scoring tools that enhance the teaching and learning process.

Steps for Creating a Rubric

Regardless of the content area, there are some basic steps that should be followed when developing a rubric. First, the specific nature of the assessment task must be identified. The task should be selected on the basis of the extent to which completing the task is expected to elicit the behaviors or evidence of understanding desired. Second, the type of rubric to be developed must be identified: Will it be holistic or analytic? Once these decisions have been made, the following steps can occur:

1. List three or four critical attributes of the performance/project. Focus on clear outcomes, which will likely be tied to the content standards. This step asks: What is important for students to learn from the activity?
2. Describe the "expected" qualities of each attribute. This will involve describing performance that goes beyond expectations, and performance that falls below expectations. This step asks: What is the standard of performance vis-à-vis each critical attribute (in other words, each cell on the rubric matrix)?

3. Evaluate the rubric. Get input from colleagues and from students. Conduct a practice run of the rubric and revise as needed. This step asks: Does the rubric capture the outcomes of importance? Does it accurately describe levels of student work? What shortcomings are evident in the rubric? How can it be improved?

The ITEA (2004) has developed a rubric template that can serve as a useful starting point for the creation of a scoring rubric (see Table 1). The finished rubric can contain as many rows as needed to capture the key objectives of the assignment. However, including too many objectives (rows) can result in a scoring rubric that is unwieldy to use. Descriptive characteristics must also be articulated for each cell in the matrix to reflect performance levels relative to each objective.

Table 1: Template for developing a performance rubric.
Note: Adapted from ITEA, 2004, p. 43.

	Not Done (0)	Beginning (1)	Developing (2)	Accomplished (3)	Exemplary (4)	Score
(Task or performance that the rubric is designed to assess identified here)						
Stated objective or "big idea"		Description of identifiable characteristics reflecting a beginning level of performance	Description of identifiable characteristics reflecting development and movement toward mastery	Description of identifiable characteristics reflecting mastery of performance	Description of identifiable characteristics reflecting the highest level of performance	

Quinlan offers a "generic" 4-point scoring key that could be modified for use with a number of scoring tools, including rubrics (p. 27). Points are awarded as follows:

 4=Advanced (in-depth understanding)—Exemplary performance or understanding; shows creativity.
 3=Proficient (general understanding)—Solid performance or understanding; the "standard."
 2=Basic (partial understanding)—Performance/understanding is emerging or developing; makes errors or demonstrates a grasp that is not thorough.
 1=Below basic (minimal understanding)—Work attempted but has serious errors; demonstrates misconceptions or weak understanding.
 0=No attempt made.

Note that 0 is always reserved for "no attempt" or "no response" (p. 26). This type of a generic scoring tool could be used in a number of ways, such as the key on an analytic rubric, or a performance checklist that has several components. This type of scoring formula takes the often-used "excellent, good, fair, poor" type of breakdown a step further by describing some of the attributes of advanced, proficient, or poor performance. It also establishes a minimum standard, while allowing for recognition of work that goes beyond what is expected.

Addressing Validity and Reliability

The challenge in creating any rubric lies first in identifying and clarifying the desired outcomes and their attributes, while at the same time not including extraneous information. In addition, there is the challenge of reducing subjectivity when describing attributes. The descriptive adjectives that are sometimes used can lead students to legitimately ask for clarification. What constitutes "some"? When is a solution "appropriate"? How many is "considerable"? Why is something "significant" or not? For this reason, it's necessary to be as descriptive as possible, to establish objective criteria when possible (e.g., "more than three spelling errors per page"), and to select the rubric components carefully. One should not waste time scoring attributes that have little to do with the overall goals of the unit or activity.

Rubrics can be evaluated for their reliability, their validity, and their utility. Reliable rubrics will result in consistent scores over time, or between multiple scorers. Valid rubrics match the stated learning objectives, as well as what was actually taught. Just as importantly, a rubric that has utility will be easy to understand and easy to use (Quinlan, 2006). Teachers will often find that they need to revise their checklists or rubrics. For example, teachers may find that trying to dissect a performance or product at too fine a level results in the inclusion of extraneous performance variables, and adds to the complexity and time needed to apply the scoring tool. They may discover that a rubric is difficult for students to understand. They may find that some critical objectives or attributes of performance have been overlooked and need to be included.

Because the essence of validity is the degree to which there is alignment between the stated learning objectives, instruction, and assessment, creating an assessment framework, like the one described in the last section, is a relatively easy test for validity that can be performed by individual teachers.

Tombari and Borich (1999) describe several types of validity, including what they call "instructional validity," which refers to the link between teaching and assessment. The two most common problems that can arise are (a) when students are not assessed on, and thus cannot demonstrate, what they have learned; and (b) when students are assessed on material that has been covered marginally or not at all.

> Teachers fall into two traps when designing assessments. The first is that they test content areas or skills that they didn't teach… The second trap occurs when the assessment places more emphasis on certain domains of knowledge or skills than was reflected during instruction… Thus, an assessment has *instructional validity* when it asks learners to do what was taught during their lessons and with the same degree of emphasis. (Tombari and Borich, 1999, p. 57)

These problems can be particularly apparent when a teacher relies on generic rubrics, or on rubrics that were developed for a context dissimilar to their own. For example, in one of my own courses I made use of a rubric developed by others for scoring student performance in a classroom debate. After the first round of use, it quickly became apparent that the rubric failed to capture some critical dimensions of performance that I had stipulated for that activity. After the third iteration of the rubric, it seemed to finally have achieved the instructional validity desired.

Lack of reliability can be a real concern with alternative assessment scoring tools. This problem may result simply from the use of a poor scoring instrument. Because alternative assessment tasks can be more time-intensive than other assessments, they may yield only a limited sample of performance or provide a too-small number of observation occasions, both of which can result in unreliable scores. Problems can also result when the task instructions are unclear, or when vague scoring standards lead to scoring imprecision (Tombari and Borich, 1999, pp. 61–63).

CLASSROOM-BASED ALTERNATIVE ASSESSMENTS

Thus far, general considerations for selecting alternative assessments and scoring tools have been discussed. In this section, I attempt to illustrate how alternative assessment decisions might be made by individual technology education teachers, with the goal of maximizing the benefits from the assessment. As was noted in Chapter 1, assessments should be conducted as part of an overall plan that includes their formative use to provide timely feedback to students and teachers. "Good assessment should be so entwined with good teaching that it becomes impossible to see where one leaves off and the other begins" (Shepard, 1997, p. 26). The purpose of any alternative assessment is to determine not only what students know, but how they use what they know. By design, alternative assessment tasks cannot be completed successfully by "good guessing" (Donovan, 2002, p. 39).

Standard 12 of the ITEA's *Standards for Technological Literacy: Content for the Study of Technology* states, "Students will develop the ability to assess the impact of products and systems" (ITEA, 2000, p. 133). Within that standard, Benchmark K for students in grades 9 to 12 states that students should be able to "synthesize data, analyze trends, and draw conclusions regarding the effect of technology on the individual, society, and the environment" (p. 138). The first job in unpacking this benchmark is to specify what learning objectives are associated with the benchmark. What kinds of data, trends, and conclusions are we interested in having students learn about? What prior knowledge do students have about these topics? What knowledge base is required to be able to synthesize data? The teacher would have to answer these questions within the context of his or her own classroom and program. Benchmark K clearly calls for a demonstration of procedural knowledge (synthesis, analysis), but will also require a demonstration of conceptual knowledge (What constitutes a trend? What evidence is required to draw defensible conclusions about an issue?).

Following the process described in Appendix H of *Measuring Progress: Assessing Students for Technological Literacy* (ITEA, 2004), we might develop a plan to assess Standard 12, Benchmark K by identifying the "big ideas" or critical attributes associated with this benchmark. Next, for each idea/attribute we would establish assessment criteria or performance expectations at, above, and below our performance targets. A draft of a rubric that could be used to assess attainment of Benchmark K is shown in Table 2. This rubric would need to be reviewed, tested, and revised. Because evidence of deep understanding of the benchmark can only be demonstrated through multiple performances in varying contexts, the same rubric could be used repeatedly to show progress over time. For example, students might be asked to examine the available evidence showing linkages between human technological activity and global warming, and then to take a position regarding those linkages or to proposed responses to the threat of global warming. In another situation, they might be asked to examine the trends regarding the size of automobiles and automobile safety. In both situations, the objectives of interest are their capacity to gather and synthesize data, and to use that data to reach and defend a conclusion.

Table 2. A Draft rubric for assessing Standard 12, Benchmark K of the *Standards for Technological Literacy*.

Given a current scenario regarding the effects of a technology, students will:	Below Target (1)	At Target (2)	Above Target (3)
Synthesize data	Locates two or fewer sources of data; does not attribute the sources, or selects sources that are of questionable accuracy; does not combine the data	Locates at least three reputable sources of data regarding the topic; attributes those sources; combines all data sources to establish a comprehensive picture of the data	Locates four or more reputable sources of data; attributes those sources; discusses conflicting pieces of data; creates a comprehensive picture of the data
Analyze trends	Does not make any predictions based on the data, or draws erroneous conclusions from the data	Describes prior effects of the technology and predicts future performance or effects based on prior performance	Achieves at target level; makes multiple predictions about future performance based on at least two different potential scenarios
Draw and defend a conclusion based on evidence gathered	Does not articulate a conclusion/opinion about the effects of the technology, or does not base conclusion on the available evidence	Articulates a conclusion/opinion about the effects of the technology and provides evidence to support that conclusion	Achieves at target level; demonstrates the ability to understand and respond to dissenting points of view

Harris, McNeill, Lizotte, Marx, and Krajcik (2006) describe a process they used to create assessments linked to science standards. The multistep process they adopted includes identifying the standard they wished to address, and unpacking that standard. Unpacking entailed clarifying the standard by expanding the primary concepts embedded within the standard and then identifying the learning performances through which students could demonstrate their ability to use the scientific ideas outlined in the standards. They addressed the issue of adequately sampling performance by first selecting learning performances that constituted multiple ways for students to demonstrate their understanding in a sequential and coherent fashion. Next, they developed assessment tools that aligned with the learning performances. The team developed what they called "base rubrics" (p. 71) that describe the components of a specific learning performance, such as scientific explanation. These base rubrics could then be modified for specific learning tasks. Finally, the team developed a "driving question" (p. 72) that would organize tasks across a series of lessons—what they called "the central organizing feature that drives students' investigations during an inquiry unit" (p. 72).

Of course, the reality of most standards-based alternative assessments is that we will not work with individual standards one by one, but will more likely link several standards in meaningful ways both in our teaching and in our assessments. Furthermore, if we are serious about the attainment of a standard (in other words, about deep understanding of core concepts), then we must "constantly work to extend that knowledge and ask about the concept in new ways" (Shepard, 1997, p. 28). We will, in other words, introduce a concept, work with it, introduce it in new contexts and with new examples, and continue to do so until a deeper understanding can be demonstrated with novel tasks. What this suggests is the need for the development of several powerful rubrics that together measure the most fundamental process skills within technology education, and their application over time.

Combining Traditional and Alternative Assessment Tools

Perhaps the most exciting, yet least tapped, resource in assessment is the use of alternative assessments designed to uncover student thinking in the performance of a task (Shepard, 1997). In technology education we often think of performance assessment as having to include the making of something, and it is the artifact that results that is assessed. However, by designing tasks that require students to show their "thinking" in the completion of a task, we can better assess student "understanding." Let us say, for example, that we want to assess students' ability to identify and defend engineering design constraints. The teacher could pose a design scenario on paper, along with representative examples of design constraints, and ask the students to select several constraints that would best enable the design of an acceptable solution. The students would be asked to describe why they selected those constraints. Based on student responses, the teacher could gain insights into the students' levels of understanding of the role played by constraints. Although this assessment task does not engage the students in the actual process of design, it is an alternative to a traditional multiple-choice test that will yield a richer source of information for the teacher.

Structuring Instruction for Effective Use of Alternative Assessments

Siegel, Hynds, Siciliano, and Nagle (2006) describe their work at the University of California-Berkeley with the Science Education for Public Understanding Program (SEPUP). The project included development of an embedded assessment system designed to measure students' thinking, decision-making, and science process skills. This embedded assessment system contained four fundamental features. First, it took a developmental perspective that focused on student learning during the school year. The same rubrics were used repeatedly during the year to document student growth on key variables. Second, to ensure curriculum alignment instructional materials and assessment tools were developed simultaneously. Third, efforts were made to establish and ensure the validity and reliability of the measures used. Finally, teachers were brought into the development process and included in creating tasks and rubrics, scoring student work, and analyzing assessment data.

Perhaps most importantly, the SEPUP approach limited the assessment effort to a small number of learning goals or "variables" (p. 93). For example, one variable was "communicating scientific information" (p. 95).

The rubric developed for this variable contained operational descriptions of the two sub-variables of interest—organization and technical aspects of communication in science. Then, rating scales for each sub-variable were created (see Table 3). The result is a simple, understandable, holistic rubric that can be used to measure a key scientific skill.

Table 3: Holistic rubric for scoring middle school science students' ability to communicate scientific information.
Source: Adapted from Siegel, Hynds, Siciliano, and Nagle, 2006, p. 95.

Score	Organization	Technical Aspects
	Response logically organizes arguments, evidence, and/or ideas related to a problem or issue. Ideas are frequently, but not always, organized in the following way: • Introduction • Explanation of procedures • Presentation of relevant evidence • Consideration of the evidence • Conclusion	Response conveys a concept or idea clearly by using the assigned medium appropriately. Possible forms of communication and ideas to examine are: • Written (e.g., report): sentence structure, grammar, spelling, neatness • Oral (e.g., presentation): enunciation, projection, and eye contact • Visual (e.g., poster): balance of light, color, size of lettering, and clarity of image
4	Accomplishes Level 3 and goes beyond in some significant way.	Accomplishes Level 3 and enhances communication in some significant way. No technical errors.
3	All parts are present and are logically organized	Presents response that is clear and easy to understand, with few minor errors.
2	Shows logical order but part is missing.	Provides an understandable response but clarity is missing in places; technical errors exist but do not prevent audience from understanding the message.
1	Lacks logical order or is missing multiple parts.	Lack of clarity and technical errors detract from audience ability to understand the message.
0	Student did not complete the task.	Student did not complete the task.

Also included in the SEPUP assessment system were assessment blueprints that gave a list of course activities along with potential assessment tasks, as well as linked test item banks. An additional tool was the identification of exemplars of student work for each scoring category, so that students would have a model for performance and teachers would have a model for scoring student work.

Based on their work with SEPUP, Siegel, Hynds, Siciliano, and Nagle (2006) have identified several recommendations for teachers to achieve a more effective use of alternative assessments. These include the need to introduce rubrics early, so that students understand their structure and their usefulness as performance guides. They recommend using the rubrics to "shift students' attention from grades to learning" (p. 101). Students can be taught to evaluate their work in relation to the criteria contained in the rubric to determine if they have included all items requested and whether they have met the standard for a desired score. These researchers also recommend establishing a system whereby rubric scores are translated into the existing letter-grade system, and where rubric scores represent just one component of an overall assessment system that might also include test and quiz scores, completion of assignments, and participation grades.

LARGE-SCALE ALTERNATIVE ASSESSMENTS: A CASE STUDY

Although the Carl D. Perkins federal legislation provides funding for career and technical education it mandates that states receiving funding, develop performance standards and measure progress toward those standards, in most cases those performance standards focus specifically on the overall CTE program and not on individual student performance. For instance, states might include in their performance standards a measure of special needs students served, which yields an overall piece of data describing numbers of students enrolled. Where large-scale efforts are made to assess "individual" students relative to established course standards, they are likely to take the form of paper-and-pencil tests of cognitive knowledge, since such tests are less costly, easier to design, less time-consuming to implement, and easier to administer and score in a secure way (Hoepfl, 2000).

In the United States, North Carolina has taken a leading role within the field of CTE in the development of standardized curriculum and state-provided support materials that include course blueprints, curriculum documents, and banks of classroom assessment items for grades 7 to 12. Course completers in nearly every CTE course at the high school level, including technology education courses, are required to take an exit exam consisting of multiple-choice questions drawn from a secure test item bank. (This system, known as VoCATS, is described more fully in Chapter 8.) In September 2004, the North Carolina Department of Public Instruction (NCDPI) initiated a pilot project to determine the feasibility of incorporating a performance assessment component as a complement to its present VoCATS system. I served as the external evaluator for that pilot project (Hoepfl, 2005).

Two or more courses from the scope and sequence model of every CTE program area were selected for inclusion in the pilot project. Teams of approximately six teachers were assembled for each course. These teachers were provided training in the development of rubrics, and then each team set about the task of creating rubrics for the course, under the direction of NCDPI consultants. The number and type of rubrics developed by these teams varied; some groups naturally gravitated toward the creation of a single, holistic rubric to evaluate a capstone project that was part of the course, while others created a series of rubrics for projects of shorter duration. The technology education teams fell into the latter category.

Baseline data were collected from all teacher participants. Analysis of the data showed that, although teachers had a good general knowledge about what alternative assessments are, only about half had made use of rubrics in their own classrooms. Three-fourths of teachers relied exclusively on the VoCATS item banks as the source of quiz and test questions for their courses. More than half expressed a desire to include more project or performance grades in their evaluation of the students (Hoepfl, 2005).

Team members created draft rubrics, tested them in their own classrooms, and then came together to revise the scoring tools. Attempts were also made to establish inter-rater reliability, and to compare student performance on the traditional versus the alternative assessment tasks.

More than 75% of the teachers said their approach to evaluation had changed as a result of participation in the pilot project. A number of teachers noted that they were using rubrics for the first time or using them more extensively than they had before. Many found that using rubrics as scoring mechanisms allowed them to be more consistent and less subjective in their grading of project work, and that grading was easier as a result of having the rubrics. These teachers found that participation in the project helped them learn more about effective grading in general. Many participants said that having clear expectations for both students and for themselves was a primary benefit of using rubrics. Several teachers expressed a belief that student performance had improved as a result (Hoepfl, 2005, p. 14).

Following a year of development work, overall response to the pilot effort was favorable, with 85% of participants expressing a desire to see the project continued. Although the feasibility project was not extended beyond the pilot year, the experience offered insights into the issues and challenges that would need to be addressed if such an effort were continued, or if the use of alternative assessment was adopted on a broader scale. These included:

- **Rubric Development.** In most cases, these teams found that it was only after extensive discussion among members that they were able to create rubrics they all agreed were of suitable quality and user-friendliness. Determining which performances to include also required extensive discussion. Recommendations made to NCDPI focused on the need to incorporate rubric development into the curriculum writing process, which would allow expectations to be clearly communicated to teachers and students at the beginning of a course. It was also recommended that rubrics be reviewed by content area experts to assess their validity relative to the curriculum standards.
- **Implementation of Performance Assessment.** Issues emerged regarding the ways that performance assessments were implemented in the classroom. Although many teams created single, holistic rubrics for scoring a comprehensive final project, teachers stressed the need to provide periodic, in-progress feedback to students rather than waiting until the end of the term. In addition to giving necessary feedback to students, this decreased the amount of time needed at the end of a term for grading. Teachers also worried about issues related to student attendance (e.g., could an assessment be recreated for students who missed the scheduled assessment day?); special needs students (e.g., should different criteria be established for these students?); availability of materials and equipment (e.g., can we afford to require performance assessments that involve the use of costly materials?); storage of projects (e.g., what kind of record can be maintained for large projects, or projects of an ephemeral nature?); and how to handle group work. Recommendations made to NCDPI focused on the need for teacher training in the use of the rubrics, which could be delivered via online modules and through professional conferences or regional workshops. Additionally, it was recommended that NCDPI establish policies to insure equitable implementation of assessments across school sites, if the alternative assessment scores were to be reported statewide.
- **Concern About How Performance Assessment Data Might be Used.** Nearly all of the participants recognized the potential for performance assessments to provide a more balanced view of student achievement

and felt it could lead to improvements in instruction. However, they wanted some assurance that the performance data would not be used in a punitive way against teachers, as some felt the traditional test data had been. Recommendations made to NCDPI to address this concern were mixed; the fundamental question on which the recommendations hinged was whether scores from the alternative assessments would be part of the state's accountability mandate, thus dictating some type of third-party scoring, or whether scores generated by teachers would be accepted as one measure of student performance, in addition to their scores on secure tests (Hoepfl, 2005, p. 4).

There are, certainly, some successful examples of large-scale alternative assessments, including assessments used for trade certification, professional licensing, and some types of advanced placements for educational programs (Hoepfl, 2000). The overarching factor in deciding whether to implement such assessments is the degree to which "security" is desired: There is an essential trade-off between insuring that scoring is objective and the cost of implementation. For many school programs, the resources required to conduct large-scale, secure, alternative assessments are simply not available. This does not mean, however, that governing bodies should overlook the use of alternative assessments in some form.

CONCLUSION

Alternative assessments hold great value in their capacity to inform the teaching and learning process. They can be powerful tools for measuring students' understanding of different types of knowledge, particularly procedural knowledge, which is not effectively assessed by traditional means. They allow students to demonstrate their knowledge in more holistic ways through an application of that knowledge. Yet alternative assessments represent just one component of what should be a balanced assessment system that will include both traditional and alternative tools.

In programs such as technology education, which emphasize both factual/conceptual and procedural knowledge, alternative assessments are an essential mechanism for measuring student capability. Teacher knowledge about all types of assessment is often limited, however, so they typically need "substantial and ongoing professional development to create valid and reliable tasks and build effective classroom assessment repertoires" (Jones,

2004, p. 586). Technology teacher education programs must include instruction on the development and use of both traditional and alternative assessments. School districts, state departments of education, and professional organizations can be important providers of assessment tools and of instruction in the use of those tools for in-service teachers.

Many educators advocate assessment reforms that feature two critical changes. First, they would like to see changes in the types of assessment tasks that are imposed on schools and teachers, to minimize the damaging effects of large-scale, high-stakes testing on the teaching and learning process. Second, they would like the "locus of assessment" to shift to the teacher, rather than toward external entities that attempt to take a one size fits all measure of learning (Darling-Hammond, Ancess, and Falk, 1995, p. 253). Significant questions remain about the viability of alternative assessments for large-scale use, particularly when the resulting scores are to be used for comparing students or districts. However, if as a society we are willing to place renewed trust in the judgment of teaching professionals, and if those professionals are trained in the use of instruction and assessment strategies that are tightly aligned with learning standards, then we might begin to realize the true power of assessment as a tool for enhancing learning and not just as a tool for holding teachers "accountable."

Brown (2004) describes how teachers trained in the Understanding by Design approach to instructional planning are able to improve their use "of a range of formal and informal assessment tools," and how the resulting picture that emerges of student performance is more like a photo album than a mere snapshot (p. 59). Well-designed assessment systems give a comprehensive picture of what works, when, and for whom.

DISCUSSION QUESTIONS

1. What are some of the important benefits of incorporating alternative assessments into an overall assessment plan?
2. For what type of technology education classroom activity would you use a scoring checklist? A holistic rubric? An analytic rubric?
3. Identify a benchmark from the *Standards for Technological Literacy* that you believe represents a foundational knowledge or skill within technology education. What critical attributes or "big ideas" would you include in a rubric designed to assess attainment of that benchmark?

4. What do you see as the primary challenges of implementing alternative assessments on a large scale? Do you think such attempts should be made? Why or why not?

REFERENCES

Anderson, L., D. Krathwohl, P. Airasian, K. Cruikshank, R. Mayer, P. Pintrich, J. Raths, and M. Wittrock, eds. (2001). *A taxonomy for learning, teaching, and assessing: A revision of Bloom's taxonomy of educational objectives.* New York: Addison Wesley Longman Inc.

Brown, J. (2004). *Making the most of understanding by design.* Alexandria, VA: Association for Supervision and Curriculum Development.

Darling-Hammond, L., J. Ancess, and B. Falk (1995). *Authentic assessment in action: Studies of schools and students at work.* New York: Teachers College Press.

Donovan, T. (2002). Illuminating understanding: Performance assessment in mathematics. *Adventures in Assessment* 14:39–42.

Harris, C., K. McNeill, D. Lizotte, R. Marx, and J. Krajcik (2006). Usable assessments for teaching science content and inquiry standards. In *Assessment in science: Practical experiences and education research,* edited by M. McMahon, P. Simmons, R. Sommers, D. DeBaets, and F. Crawley, 67–87. Arlington, VA: National Science Teachers Association Press.

Hoepfl, M. (2000). Large-scale authentic assessment. In *Using authentic assessment in vocational education,* edited by R. Custer, 49–67. Columbus, OH: ERIC Clearinghouse on Adult, Career, and Vocational Education.

——— (July 2005). NCDPI Career-technical education performance assessment feasibility project: External evaluation final report. Evaluation report presented to the North Carolina Department of Public Instruction, Raleigh, NC.

International Technology Education Association (2000). *Standards for technological literacy: Content for the study of technology.* Reston, VA: International Technology Education Association.

——— (2004). *Measuring progress: Assessing students for technological literacy.* Reston, VA: International Technology Education Association.

Jones, K. (2004). A balanced school accountability model: An alternative to high-stakes testing. *Phi Delta Kappan* 87:584–590.

Lederman, L. M., and R. A. Burnstein (2006). Alternative approaches to high-stakes testing. *Phi Delta Kappan* 87:429–432.

Missouri Department of Elementary and Secondary Education (1994). *Technology education performance-based education implementation handbook.* Columbia, MO: University of Missouri-Columbia Instructional Materials Laboratory.

National Research Council (2001). *Knowing what students know: The science and design of educational assessment,* edited by J. Pellegrino, N. Chudowsky, and R. Glaser, Washington, DC: National Academies Press.

——— (2002). *Investigating the influence of standards: A framework for research in mathematics, science, and technology education.* Washington, DC: National Academies Press.

Quinlan, A. (2006). *A complete guide to rubrics: Assessment made easy for teachers, K–college.* Lanham, MD: Rowman & Littlefield Education.

Scott, J. (2000). Authentic assessment tools. In *Using authentic assessment in vocational education,* edited by R. Custer, 33–48. Columbus, OH: ERIC Clearinghouse on Adult, Career, and Vocational Education.

Siegel, M., P. Hynds, M. Siciliano, and B. Nagle (2006). Using rubrics to foster meaningful learning. In *Assessment in science: Practical experiences and education research,* edited by M. McMahon, P. Simmons, R. Sommers, D. DeBaets, and F. Crawley, 89–106. Arlington, VA: National Science Teachers Association Press.

Shepard, L. A. (1997). *Measuring achievement: What does it mean to test for robust understanding?* Princeton, NJ: Educational Testing Service.

Tombari, M. L., and G. D. Borich (1999). *Authentic assessment in the classroom: Applications and practice.* Upper Saddle River, NJ: Merrill/Prentice-Hall Inc.

Walberg, H. J., G. D. Haertel, and S. Gerlach-Downie (1994). *Assessment reform: Challenges and opportunities.* Bloomington, IN: Phi Delta Kappa Educational Foundation.

Wiggins, G., and J. McTighe (2001). *Understanding by design.* Upper Saddle River, NJ: Prentice-Hall Inc.

Assessing Students with Disabilities in Technology Education

Chapter 5

Phillip L. Cardon
Eastern Michigan University

INTRODUCTION

As assessment in education continues to evolve, the field of technology education must also make changes in order to match those assessments. The need to assess students with disabilities in the technology education classroom is a part of these assessments. This chapter will discuss the characteristics of assessing students with disabilities, along with traditional and alternative assessment methods. It will also discuss how legislation and standards influence the assessment of students with disabilities, and how to communicate the assessment needs of students with disabilities to school administrators, teachers, and parents.

Background of Assessing Students with Disabilities in Technology Education

Throughout the history of the field, technology education assessment has been related to performance regarding subject matter and related laboratory skills (Kimeldorf, 1984). Concurrently, much of the dialogue has been related to the assessment of the "typical" student; that is to say, the nondisabled student. However, since the passage of the Education for All Handicapped Children Act of 1975 (also known as Public Law [PL] 94–142), assessment of children with disabilities has been a requirement. This act requires "a free and appropriate public education and related services designed to meet [the] unique needs" of each student (Kimeldorf, 1984, p. 6). PL 94–142 defines students with disabilities as those with visual or hearing disabilities, mental capacity reduction, serious emotional or orthopedic difficulties, or multiple disabilities. Later, under the Individuals with Disabilities Education Act of 1990, students with brain injuries or autism were added to the list (Podemski et al., 1995). Although PL 94–142 gave students with disabilities an opportunity to participate in regular educational activities, it posed a reforming but positive challenge for technology educators and administrators: how to incorporate students with disabilities into all phases of technology education and laboratory activities.

Importance of Assessing Students with Disabilities in Technology Education

Since a portion of the approximately 6.29 million students in the United States with disabilities may be enrolled in secondary technology education courses, technology education personnel and school administrators should develop and follow an assessment plan that includes Individual Education Plans (IEPs) for students with disabilities (National Center for Education Statistics, 2002).

The term *assessment* as it relates to students with disabilities is defined as "the systematic process of gathering educationally relevant information in order to make legal and instructional decisions about the provision of special services to students with disabilities" (McLoughlin and Lewis, 1994, p. 601). Some considerations relative to the assessment of students with disabilities include the characteristics of quality assessments, traditional versus alternative assessments, psychomotor skills and safety, legislative issues, and the relationship with the *Standards for Technological Literacy* (International Technology Education Association [ITEA], 2000) and its assessment companion, *Advancing Excellence in Technological Literacy* (ITEA, 2003). These play an important role in assessment in the field of technology education and need to be communicated to school administrators, teachers, and parents.

CONSIDERATIONS WHEN ASSESSING STUDENTS WITH DISABILITIES IN TECHNOLOGY EDUCATION

Multi-Factored Assessment of Students with Disabilities

Multi-factored assessment refers to the "assessment and evaluation of a handicapped child with a variety of test instruments and observation procedures" (Heward and Orlansky, 2002, p. GL–10). Some of the questions to be considered in the multi-factored assessment of students with disabilities include the methods of observation and interview to be used, student learning-environment interactions, effective interventions, and family characteristics and influences (McNamara and Hollinger, 2003).

A study performed by McNamara and Hollinger (2003) determined multi-factored assessments to be very beneficial in determining intervention strategies for students with disabilities. In addition, they found that problem solving in hands-on assessment models better served students with disabilities and met school goals. Technology education classrooms would appear to fit well with a multi-factor approach, since problem-solving procedures may provide students with disabilities more effective hands-on assessments than traditional core-content procedures do.

Validity in Assessing Students with Disabilities

Since students with disabilities are included in the general classroom and assessed using each state's general academic instrument, this assessment needs to be performed without interference from the disabilities in order to obtain a true academic score. Validity is evident when a test assesses what it is supposed to assess and nothing more. Traditional measures of assessment validity, such as content, predictive, and face validity, must also be present when assessing students with disabilities (Power, 2000).

According to Koretz and Barton (2003), for assessment of students with disabilities to be valid, issues of identification and classification of disabilities, appropriate accommodations, evaluation of proficiency, bias, and instrument design need to be addressed. Technology educators must evaluate curriculum materials, assessment instruments, procedures, and environments, and modify them to match the needs of students with disabilities. This could include the modification of laboratory equipment and tools or the use of larger text fonts and/or Braille documents. For example, when assessing a disabled student's knowledge and understanding of drawing concepts, he or she may need modified books and reading materials or adapted drawing tools. Modified test questions may include simple wording and picture examples in order to help students with disabilities better understand the question. Modifications to tools and equipment for evaluation purposes should be included so that the student's disabilities do not interfere with performance on the evaluation. If a disability prevents a student from using his right arm, then the laboratory assessment activity should be modified to allow the student to perform the assessment without using his right arm, in order not to compromise his score or rating.

Bias-Free Assessment of Students with Disabilities

According to Popham (2002), bias "refers to qualities of an assessment instrument that offend or unfairly penalize a group of students because of the students' gender, ethnicity, socioeconomic status, religion, or other such group-defining characteristics" (p. 73). This would also include disabling characteristics.

One way to help ensure the bias-free assessment of students with disabilities is to work with a bias review panel. The panel would review assessment materials and provide feedback regarding possible bias-generating items (Popham, 2002). For example, bias may occur in a technology education laboratory setting when an assessment requires a student to stand to use a scroll saw. A student who cannot stand would not be provided an adequate assessment opportunity, making it difficult for him or her to perform as expected. The assessment bias may be removed in this example by providing a scroll saw station suitable for someone who needs to sit in a wheelchair. Regardless of whether a panel is used, technology education teachers should be cognizant of, and work to prevent, bias in assessment.

Laboratory Safety Assessment of Students with Disabilities

Because technology education curriculum is usually connected to a laboratory setting, a discussion relative to the laboratory safety assessment of students with disabilities is important. Two primary considerations when assessing laboratory safety are (a) the physical abilities of the student, and (b) lab characteristics that may pose safety concerns. If an instructor considers a student with disabilities to be a safety risk to himself, or to the rest of the students in the class, then the teacher should consider counseling the student with disabilities in another class setting (Storm, 1993). Determination of safety risks can begin with an examination of the student's individualized education plan (IEP), and should include a safety assessment of the student in the laboratory setting, as well as a discussion with counselors and/or parents regarding potential safety risks.

A number of measures can be taken to evaluate the safety of students with disabilities in the technology education program (Kimeldorf, 1984). These include modifying safety tests, using aides or peer tutors to monitor safety behavior, and carefully documenting observed safety problems.

MODIFYING TRADITIONAL ASSESSMENT TOOLS FOR STUDENTS WITH DISABILITIES IN TECHNOLOGY EDUCATION

Some of the traditional assessment methods used in technology education to assess students in general—and students with disabilities, in particular—include the response assessment method and the computer-based assessment method. Both of these will be discussed in this section.

Response Assessment of Students with Disabilities

Because assessment is critical to mapping the learning progress of students with disabilities in technology education, it is important to understand the various formats of assessment instruments, and their advantages and disadvantages (Goh, 2004; Johnson and Johnson, 2002). Some common formats include objective tests (e.g., multiple choice, true-false, matching, short answer or completion, and interpretive), essay tests (e.g., essays and short essays), and cooperative learning tests (e.g., group discussion tests).

The following are some ways in which objective, essay, and cooperative learning tests may be modified to become better suited for students with disabilities. When writing objective tests, try to use simple words, and capitalize words that are important for students to remember. With multiple-choice questions, do not use more than three choices per question. Provide a word bank for short-answer and essay questions. Whenever possible, use interactive assessments such as group tests in the laboratory, or discussion groups in the classroom (Buffer and Scott, 1986; Sarkees-Wircenski and Scott, 1995). An example might be for the instructor to have a group of students work on a problem and write down the things they did and the topics they discussed while they tried to work through the problem. During the activity, the instructor should check to see how the students are progressing and discuss the results with the group following the assessment activity.

Computer-Based Assessment of Students with Disabilities

As with mainstream education, the computer has had an influence on the assessment of students with disabilities. The computer is seen by some as enabling "more reliable, cost-effective, and sophisticated" assessments than traditional practices (Power, 2000, p. 205).

As part of most modular technology education curriculum models, and in many nonmodular labs, as well, the computer is an excellent tool for assessing students with disabilities. Utilizing assistive technology software, students can have texts read to them at the speed they choose. With modifications to the input devices, students can take self-paced tests on the computer, allowing them to be assessed for comprehension and recall (Goh, 2004). Computer-based simulations might allow for testing of knowledge and skills that might be impractical or dangerous in an actual lab setting.

ALTERNATIVE ASSESSMENTS USED FOR ASSESSING STUDENTS WITH DISABILITIES IN TECHNOLOGY EDUCATION

Although alternative assessments are commonly used in technology education, the use of alternative assessments for evaluating students with disabilities, specifically, in technology education is critical to the success of these students (Sarkees-Wircenski and Scott, 1995). Some of the alternative assessment methods used in technology education to assess these students include portfolio assessments, affective assessments, group-oriented assessments, other so-called "authentic" assessments, and psychomotor skills assessments (Tindal et al., 2003; Towles et al., 2003).

Portfolio Assessment of Students with Disabilities in Technology Education

As an alternative to traditional assessment methods, portfolio assessment evaluates the knowledge and skills of students through a review of their work, and is defined as "the procedure used to plan, collect, and analyze the multiple sources of data that reflect a student's accomplishments" (Goh, 2004, pp. 149–150). These assessments may include completed assignments, quizzes, tests, and other materials that form an organized document, which gives a picture of the student's abilities.

Portfolio assessment for students with disabilities in technology education might be used in a manufacturing course that teaches problem-solving procedures, equipment safety, integration, and teamwork. For example, when the students are assigned to learn about a manufacturing enterprise, the instructor can have students make an audio or videotape describing the enterprise. The students might draw pictures describing the manufacturing activity performed by the enterprise and then actually participate in the manufacturing activity. A summary paper, video, or audio recording describing their experience would conclude the activity. All of these products would become part of the portfolio assessment.

When evaluating the portfolio of students with disabilities in technology education, teachers should assess students according to their ability to understand content and perform tasks involved in the assignments and activities. The portfolio assessment of students with disabilities might be different from the portfolio assessment of regular students in that the limitations of the students with disabilities need to be considered and accounted for when using scoring mechanisms (Kossar, 2003).

Affective Assessment of Students with Disabilities in Technology Education

Affective assessments are defined as "those that deal with a student's attitudes and values, such as the student's self-esteem, risk-taking tendencies, or attitudes toward learning" (Popham, 2002, p. 99). Technology education programs tend to have a positive affective result on students with disabilities (Buffer and Scott, 1986). Some of these positive affective results include the satisfaction students feel about their completed work at the end of a technology education group activity, and the appearance of a more positive attitude toward learning in an integrated environment that enhances understanding (Popham, 2002). Teachers may assess the affective results of students with disabilities through asking simple questions about how the students felt about the activity. As educators help students to see the value of learning and assess that value or affect, they will be able to better assist students to learn by focusing on the affective skills that will best help each student (Towles et al., 2003).

Group-Oriented Assessment of Students with Disabilities

Through student collaboration, the group-oriented learning experience has been known to not only improve social skills, but enhance learning. When performing a group-oriented assessment, it is important to determine if the purpose of the assessment is to measure the knowledge of students, or their ability to collaborate (Webb, 1997). Clarity about the focus of the assessment and subsequent attention to the design of assessment tools that measure the desired abilities will lead to greater validity of group-oriented assessments.

Assessing Psychomotor Skills of Students with Disabilities

Psychomotor skills are an important part of teaching students with disabilities in technology education (Sarkees-Wircenski and Scott, 1995). Assessments in the psychomotor skills area have traditionally taken the form of tasks or checklists. Most of these assessment tools focus on the student's ability to display specific skills or complete assigned tasks such as keyboarding skills.

To assess psychomotor skills of students with disabilities, a preliminary skills assessment should be performed to determine the student's capacity to complete a required activity, given the requisite training. If the student is found to be lacking psychomotor abilities in a particular area, the instructor has two options. Either the instructor may provide supplemental assistance, so the student can acquire the minimum amount of psychomotor abilities, or the instructor may remove the student from the group. Accommodations may be necessary to ensure that the student will have a fair assessment in technology education.

LEGISLATION AFFECTING ASSESSMENT OF STUDENTS WITH DISABILITIES IN TECHNOLOGY EDUCATION

In addition to the Americans with Disabilities Act (ADA) of 1990 and the Individuals with Disabilities Education Act (IDEA of 1975), several pieces of legislation that affect assessment of students with disabilities in technology education include the No Child Left Behind Act of 2001 and the Carl D. Perkins Career and Technology Education Act (1984, 1990, 1998). Each of these pieces of legislation will be discussed in relation to its impact on the assessment of students with disabilities in technology education.

No Child Left Behind Act and Its Effect on Assessing Students with Disabilities

Since the ratification of the No Child Left Behind (NCLB) Act in 2001, schools and programs have made changes to accommodate NCLB. This legislation has affected all aspects of the education process in K–12 schools, including the education of students with disabilities in technology education. The integrative properties of the technology education curriculum allow it to be a significant contributor to the development of student abilities in mathematics, science, and social science (LaPorte and Sanders, 1995). As more frequent testing within a greater range of subject areas is implemented as a result of NCLB, measures that enhance student learning in the core subject areas will become more important. Some preliminary studies show that participation in technology education programs helps students perform better on assessments in core subject areas (Ernst, Taylor, and Peterson, 2005; Satchwell and Loepp, 2002).

Americans with Disabilities Act and Its Effect on Assessing Students with Disabilities

The Americans with Disabilities Act (ADA) of 1990 (PL 101-336) gave civil rights protection to people with disabilities, which is similar to what the Civil Rights Act of 1974 did for people in the areas of gender, race, religion, and national origin (Heward, 2003; Sarkees-Wircenski and Scott, 1995). The ADA also protected the rights of students with disabilities in education. It required administrators and educators to be aware of circumstances that inhibit the equitable education of students with disabilities (Sarkees-Wircenski and Scott, 1995).

The ADA is important to the assessment of students with disabilities in technology education. Technology education teachers need to be aware of circumstances that prevent students from receiving equal learning opportunities and assessments in classrooms and laboratories. For example, physical barriers to wheelchair access should be removed, and a student's disability should not inhibit the use of equipment and tools. Alternative assessments may be considered, and proper accommodations may be given to the student. As teachers follow the guidelines set forth in the ADA, they may help to ensure the equitable assessment of students with disabilities in technology education classroom and laboratories.

Individuals with Disabilities Education Act and Its Effect on Assessing Students with Disabilities

In 1983, the Education for All Handicapped Children's Act (EAHCA) was amended by PL 98–199, which gave provisions for the education of students with disabilities during preschool years, and assisted students making the transition to adulthood. A further amendment in 1986 (PL 99–457) gave additional provisions to children with disabilities from birth to pre-school age (Heward, 2003). An amendment in 1990 (PL 100–476) changed the name of the EAHCA to the Individuals with Disabilities Education Act (IDEA; PL 101-476) and added traumatic brain injury and autism to the list of supported categories. This act also required transition services to be in the student's IEP prior to age 16 (Sarkees-Wircenski and Scott, 1995).

The most recent amendment of the IDEA (1997; PL 105–17) greatly expanded the requirements of educational provisions for students with disabilities. These requirements included more parent participation, IEP modifications, and general education opportunities (Heward, 2003, p. 33).

As with the ADA, the IDEA has an effect on the assessment of students with disabilities in technology education. Technology education teachers should encourage parents to participate in decisions about the student's education and provide input regarding the best instruction and assessment methods for the student. This will help the technology education teacher make the proper accommodations for the student with disabilities. Technology education teachers should ensure that the student's IEP is being followed and regularly reviewed by school personnel and parents. In this way, the teacher can follow the guidelines set forth by the IDEA regarding assessing students with disabilities.

Carl D. Perkins Career and Technology Education Act and Its Effect on Assessing Students with Disabilities

The Carl D. Perkins Vocational Education Act (PL 98–524) of 1984 was established to help people obtain the skills they need for employment and assist with national economic development efforts. It also addressed the issue of access to education by handicapped and disabled persons (Sarkees-Wircenski and Scott, 1995).

In 1990, this legislation was amended by the Carl D. Perkins Vocational and Applied Technology Education Act (PL 101–392). This legislation addressed the integration of academics with vocational education, and provided additional access to technical skills preparation by special needs populations (Sarkees-Wircenski and Scott, 1995). It also ensured that students with special needs would have equal access to education recruitment, enrollment, and placement activities, among other things. The needs of students with disabilities were similarly addressed in the 1998 Perkins Act amendment (PL 105–332). The Perkins legislation also requires that proper assessments be given to these students (Sarkees-Wircenski and Scott, 1995).

The mandate contained in all of these pieces of legislation is clear: All teachers, including technology education teachers, must work to recognize and accommodate the needs of students with disabilities in their classrooms and laboratories by modifying their curriculum, instruction, and assessment.

COMMUNICATING ASSESSMENT NEEDS OF STUDENTS WITH DISABILITIES IN TECHNOLOGY EDUCATION

In order for the assessment of students with disabilities in technology education to be successful, the needs of the students must be conveyed to school administrators, teachers, and parents to encourage a team effort for assessment. One of the primary avenues for informing this team of the needs of students with disabilities is through the use of an IEP. This document, required for all students with disabilities, is developed by a team that includes (but is not limited to) a school administrator, the teacher, and a parent or guardian (Heward, 2003).

Technology education teachers can use the IEP to assist them in preparing lessons and activities that include the student with disabilities. The IEP will describe the educational level of the student, educational goals, specific required services, and assessment procedures (Sarkees-Wircenski and Scott, 1995). The IEP will provide guidance for assessing students in order to help them be successful. Other resources that may be available to teachers of students with disabilities include tutoring programs, speech software and Braille document creation, education outreach programs, and grants. For more information about these and other resources, contact your school district or state department of education.

CONCLUSION

Through legislation such as the Individuals with Disabilities Education Acts of 1975 and 1997; the Carl D. Perkins Career and Technology Education Acts of 1984, 1990, and 1998; the Americans with Disabilities Act of 1990; and the No Child Left Behind Act of 2001, the rights of students with disabilities have been significantly expanded. With these opportunities and rights come many challenges in education, technology education, and its role in providing equitable educational services to students with disabilities. Some of these challenges include ensuring the use of valid, reliable, and bias-free assessments of these students.

To assist with the assessment of students with disabilities in technology education, ITEA facilitated development of the *Standards for Technological Literacy* (ITEA, 2000) and *Advancing Excellence in Technological Literacy* (ITEA, 2003). Together, these documents have helped to establish a foundation for the content in the field and the assessment of students—including students with disabilities—in technology education.

There have been a number of concerted efforts to advance the assessment of students with disabilities in technology education, but the work is far from complete. With the inclusion of students with disabilities in the technology education classroom, it is the responsibility of the technology education teacher to know and understand the rights of these students and how to provide appropriate and equitable learning experiences and assessments in the technology education classroom and laboratory. It is important that technology education teachers read and understand the laws governing the assessment of students with disabilities and stay abreast of future legislation affecting the rights of these students. It is hoped that technology education teachers will be able to understand and use the resources referred to in this chapter. In this way, teachers may be empowered and feel confident in the assessment of students with disabilities in technology education.

DISCUSSION QUESTIONS

1. What are some curriculum assessment issues related to teaching students with disabilities in technology education classrooms?
2. What are several ways a technology education teacher can ensure that students with disabilities are properly assessed in technology education classrooms?
3. How might instructional technologies, including computers, help assess students with disabilities in technology education classrooms?
4. What pieces of federal legislation have influenced the assessment of students with disabilities in technology education classrooms?
5. How might the STL and/or AETL be used to inform the process of assessing students with disabilities in technology education classrooms?

REFERENCES

Buffer, J. J., and M. L. Scott (1986). *Special needs guide for technology education.* Reston, VA: International Technology Education Association.

Custer, R. L., J. Schell, B. D. McAlister, J. L. Scott, and M. Hoepfl (2000). *Using authentic assessment in vocational education.* Columbus, OH: ERIC Clearinghouse.

Ernst, J. V., J. S. Taylor, and R. E. Peterson (2005). Tech-know: Integrating engaging activities through standards-based learning. *The Technology Teacher* 65 (2): 15–18.

Goh, D. S. (2004). *Assessment accommodations for diverse learners.* Boston, MA: Pearson.

Heward, W. L. (2003). *Exceptional children: An introduction to special education.* Upper Saddle River, NJ: Pearson Education Inc.

Heward, W. L., and M. D. Orlansky (2002). *Exceptional children.* New York: Macmillan Publishing Company.

International Technology Education Association (2000). *Standards for technological literacy: Content for the study of technology.* Reston, VA: International Technology Education Association.

———— (2003). *Advancing excellence in technological literacy: Student assessment, professional development, and program standards.* Reston, VA: International Technology Education Association.

Johnson, D. W., and R. T. Johnson (2002). *Meaningful assessment: A manageable and cooperative process.* Boston, MA: Allyn & Bacon.

Kimeldorf, M. R. (1984). *Special needs in technology education.* Worcester, MA: Davis Publications Inc.

Koretz, D. M., and K. Barton (2003). Assessing students with disabilities: Issues and evidence. *CSE Technical Report.* (ERIC Document Reproduction Service No. ED 474869).

Kossar, K. (2003). Graduate practicum—Special education: Assessment through portfolio development. *Teacher Education and Special Education* 26 (2): 145–49.

LaPorte, J. E., and M. E. Sanders (1995). Integrating technology, science, and mathematics education. In *Foundations of technology education,* edited by G. E. Martin, 179–220. Peoria, IL: Glencoe/McGraw-Hill.

McLoughlin, J. A., and R. B. Lewis (1994). *Assessing special students.* 4th ed. New York: Merrill.

McNamara, K., and C. Hollinger (2003). Intervention-based assessment: Evaluation rates and eligibility findings. *Exceptional Children* 69 (2): 181–93.

National Center for Education Statistics (2002). *Digest of education statistics 2001, Chapter 2: Elementary and secondary education.* http://nces.ed.gov/pubs2003/2003060b.pdf#search='secondary%20student%20disabled%20statistics (accessed February 27, 2005).

Podemski, R. S., G. E. Marsh II, T. E. Smith, and B. J. Price (1995). *Comprehensive administration of special education.* 2nd ed. Englewood Cliffs, NJ: Prentice Hall.

Popham, W. J. (2002). *Classroom assessment: What teachers need to know.* Boston, MA: Allyn & Bacon.

Power, P. W. (2000). *A guide to vocational assessment.* Austin, TX: Pro-Ed.

Satchwell, R.E., and F. L. Loepp (2002). Designing and implementing an integrated mathematics, science, and technology curriculum for the middle school. Journal of Industrial Teacher Education 39 (3): 41–66.

Sarkees-Wircenski, M. D., and J. L. Scott (1995). *Vocational special needs.* Homewood, IL: American Technical Publishers Inc.

Storm, G. (1993). *Managing the occupational education laboratory.* Ann Arbor, MI: Prakken Publications.

Tindal, G., M. McDonald, M. Tedesco, A. Glasgow, P. Almond, L. Crawford, and K. Hollenbeck (2003). Alternative assessments in reading and math: Development and validation for students with significant disabilities. *Exceptional Children* 69(4): 481–494.

Towles, E. A., B. Garrett, B. Burdette, and M. Burdge (2003). *What are the consequences? Validation of large-scale alternative assessment systems and their influence on instruction.* (ERIC Document Reproduction Service No. ED 478482)

Webb, N. M. (1997). Assessing students in small collaborative groups. *Theory Into Practice* 36(4): 205–213.

Data Analysis Techniques

Chapter 6

Michael R. Lindstrom
Anoka-Hennepin Schools

CHAPTER OVERVIEW

This chapter addresses a variety of topics related to analyzing assessment data. Topics include individual and group analysis as well as numeric treatments of declarative and procedural data. While the chapter stops short of outlining the statistical processes used in analyzing data (many commonly accessible texts address those techniques), the considerations in preparing data for analysis as well as options for focusing the analysis within technology education classrooms are addressed. Also included are the unique opportunities data analysis provides to improve assessments, learning, and instruction as well as assessing the effectiveness of local, state, and national programs.

The goals of the chapter are to assist the reader in
- understanding the differences between individual and group data analysis and their respective advantages/disadvantages.
- understanding the types of data and data applications that are unique to technology education classrooms.
- exploring a variety of options for gathering data and their implications for analysis.
- understanding the various techniques for preparing data.
- understanding the various purposes of performing data analysis.
- understanding a variety of considerations related to reporting the analysis of data.

IMPORTANCE OF DATA ANALYSIS

The technology education classroom is unique in terms of the types of data gathered. While traditional test scores will comprise a portion of the data, other aspects will include performance assessment data, laboratory and safety data, and, in the case of computer-managed curricula such as modular laboratories, management system data. Each of these needs to be analyzed individually, combined with other aspects as appropriate, and ultimately integrated into either a holistic course grade for students or an improvement plan for instruction or programs.

Performance assessment data are derived from the products or projects on which students work. The data produced by these assignments is likely to be subjective in nature, requiring the use of scoring rubrics. In the author's experience, teachers also feel that the performance portion of technology education courses needs to be much more heavily weighted than declarative test data. This makes data analysis within technology education courses unique from virtually all traditional academic courses.

At first glance, the data analyzed in technology education classrooms may appear to have the same applications as data from other courses; after all, reporting student results and performing analysis to improve instruction are critical attributes of assessment in any classroom. However, in K–12 school systems, technology education courses and programs are typically elective in nature, and data analysis can serve the political function of providing evidence that these courses are effective and viable portions of the school curriculum. Viability is an issue at the post-secondary level as well, and technology education programs can be supported by analyzing and documenting outcomes such as placement data, student surveys, and employment trends. Data may also be needed to justify the greater budgetary demands of technology education programs, or to monitor the success of the safety aspects of the courses. Finally, data analysis can be used to support the need for, and success of, embedding technology concepts in traditional core classes such as science and social studies.

ALIGNING ANALYSIS WITH PURPOSE

In his paper "Pitfalls of Data Analysis," Helberg (1995) describes the primary purposes of data analysis as being able to accurately describe a group's performance, make inferences about a group, or predict a group's future performance. If these purposes are to be accomplished, then any analysis must eliminate bias, avoid methodological errors, and accurately interpret and communicate results. Furthermore, the insights gained through data analysis should ultimately be used to improve instruction and programs.

To avoid bias, data gathered for analysis must: (a) truly represent the target group (i.e., employ proper sampling techniques); (b) be collected in a way that ensures the sample actually does represent a normally distributed larger population; and (c) be valid (i.e., verify that the learning being assessed resulted from the intervention of instruction within a particular course). In addition to expected differences between individual students, data from students in a given class will also be dependent on the instructor and other classes scheduled at the same time that compete for particular types of students.

The concept of validity applies to both the construction of assessment instruments and to the analysis of results. With regard to data analysis, the validity question basically asks, "Do the data accurately reflect what it is we wish to analyze?" For example, if analysis is being performed on safety data, and the technology education teacher only keeps records of accidents that require a trip to the school nurse or a doctor, then an incomplete picture of the safety program will result. When looking at the effectiveness of safety instruction, validity would be increased if data were maintained and analyzed for all of the incidents that occurred, including those that simply required a trip to the classroom first-aid kit.

The purpose of formative data analysis is typically to provide feedback on the learning and instructional processes occurring in the classroom, thereby assisting the teacher in timely interventions and adjustments in instructional pacing and strategies. Analysis of summative data is better suited for reporting student results (grading) and determining course/program effectiveness. Growth data (a combination of the two) can inform both purposes. Data collection, data analysis, and reporting strategies will necessarily vary, depending on the nature of a particular assessment.

Too often, data analysis is focused on results from a single, summative test. Although this approach may provide useful information about a student's status relative to a standard or fixed measure, it provides little information about learning as a result of participation in a particular course or about the effectiveness of teaching within the course. Alternatively, measuring growth alone (i.e., change in achievement over time) will not necessarily provide information about the final level of competence students have obtained. An ideal approach to data analysis would provide information about both aspects of learning in a classroom setting: analysis of various aspects of growth within a course over time, and a summative analysis that shows achievement relative to a standard.

Assessing Declarative and Procedural Knowledge

Declarative knowledge is learning that can be withdrawn from memory in response to questions. Asking students to define *technology* or list the four elements of a typical "systems model" would be declarative examples. Declarative knowledge is critical to student success and is often considered basic foundational or background knowledge. It is most appropriately assessed with a paper-and-pencil test. Unfortunately, procedural knowledge (the ability to "do" something) is difficult to assess via pencil and paper and is usually assessed with performance assessments that incorporate observing students as they apply knowledge, or by evaluating student products.

A comprehensive assessment system intended to verify one's readiness to perform oxyacetylene welding, for example, would ask questions about gas pressures and names of equipment parts (declarative knowledge) and then ask the student to produce sample welds (procedural knowledge). Both aspects are critical to assessment in technology education, and the proportion of assessments that are procedural (performance-based) is likely to be significantly higher in technology education than in most other content areas.

Assuring the Validity of Technology Education Assessments

Teachers may establish validity of declarative items by
- reviewing the focus of each question and making sure it is aligned with the intent of the standards/domain being assessed, as well as ensuring that questions do not focus on irrelevant knowledge (content validity).
- reviewing the terms used in questions and aligning them with the standards/domain being assessed (face validity).
- reviewing the assessment as a whole to ensure that all of the domains of knowledge intended to be measured have been systematically included (content validity).

- asking another teacher or field expert to perform the steps above.
- purchasing or using a set of items prescreened for validity on a topic, such as items developed as part of a state assessment system.

If these types of steps have not been taken, the likelihood of some or all items being valid is relatively low. The following example applies these steps to a declarative item focused on the "Nature of Technology" (International Technology Education Association, 2000). The question a teacher might ask of students is, "Is the rate of technological development increasing or decreasing?" Reviewing the key terms found in the question (rate, technological, development, increasing/decreasing) will show that they are the exact terms found in the *Standards for Technological Literacy* (STL), thus supporting face validity. The concept of an increasing technological rate is an important concept within the standards as well, providing content validity—assuming that this concept was actually addressed as part of the information presented in the course.

Content validity is further established by reviewing each question individually to ensure that the item focus is, in fact, part of the domain of study/standards, and that no significant concept or critical attribute of that domain has been omitted. A question asking, "Is the rate of technological development increasing exponentially?" could easily fail the content validity test if the mathematical concept of exponential growth was not addressed in the course content. Finally, if a comprehensive assessment on "design" from the STL failed to ask questions about the role of troubleshooting, research and development, or invention in problem solving, then essential understandings about key concepts in design have been omitted and content validity has been reduced. Asking experts or other technology teachers to review the assessment is one way to identify such gaps.

Stiggins, Arter, Chappuis, and Chappuis (2004) suggest constructing an assessment matrix that lays out a plan for the strands or topics addressed by a test and the number/type of items within each strand. A minimum of six to eight items are usually needed to adequately address a strand, which will typically limit the number of strands addressed by a summative assessment to less than 10. Strands may also weight items in point value or quantity to the proportion the value of that strand has in relation to others.

Establishing the validity of procedural items (performance tasks) is similar, though slightly more complex, and would include
- reviewing the task directions to ensure that the terms used are consistent with the standards/domain being assessed (face validity).
- reviewing the task directions to ensure that students must demonstrate the knowledge and skills required in the standard/domain being assessed, and are not dependent on irrelevant knowledge and skills (content validity).
- reviewing the tasks as a whole to ensure that all of the domains of knowledge we wish to measure are being measured (content validity).
- asking another teacher or field expert to review the assessment tasks.
- purchasing a set of procedural task items prescreened for validity.

Since there will typically be fewer procedural items than declarative items, because procedural items generally require more time to complete, these items must also be analyzed for their robustness or ability to address a broad range of the construct being measured. An alternative to a single, robust performance, suggested by Wiggins and McTighe (2005), would be to use scores from a variety of student work samples gathered during a period of time and combine them with a summative assessment. Either way, a truly valid performance assessment would address the standard/domain being assessed in its entirety, and would not include information extraneous to the standard or domain.

Consider the following assignment as an example of a performance task focused on design: Design an object that will assist elderly people in opening child-proof medicine bottles, and document the design process used. This task description would need additional details, and face validity would be verified by ensuring that the terminology in those details is directly taken from the course standards and classroom instruction. Since design and design processes are part of the STL, content validity will exist as long as the task details and scoring criteria reflect the essence of what is contained in the standards relative to design, and as long as those aspects have been addressed within the course.

TECHNIQUES FOR DATA GATHERING

Obtaining Data About Individual Student Growth

To measure student growth, a baseline must first be established. This requires a pretest of some type, or preexisting data from a previous course. Determining a growth score will also require a post-test or summative assessment that has a high degree of alignment with the pretest (in other words, identical or highly correlated questions). If the number of items and overall point value of both the pretest and post-test are identical, then simple subtraction (post-test minus pretest score) will produce a growth score.

Growth scores from a valid assessment will inform the teacher about the achievement of individual students and about the effectiveness of instruction. Aggregating data across a number of identical classes, either over a number of years with a single teacher, or across multiple teachers' classes, will allow the creation of norms for references of "typical growth." Effectiveness of individual teachers can then be gauged by determining how the mean growth of a class compares to typical growth means (deviation from the typical growth means). Hierarchical linear modeling (HLM) and other data techniques can be used to look at teacher effect (in a sense, the "value added" by a teacher) over a number of years. This type of multiyear analysis is critical to providing accurate measurements of teacher effectiveness, since data from a single year or a single class is subject to the cohort effect, where skewed scores may result when classroom populations are comprised of unusually high numbers of either talented or challenged students.

Obtaining Data About Groups

"If we wish to maximize student achievement in the U.S., we must pay far greater attention to the improvement of classroom assessment. Both assessment *of learning* and assessment *for learning* are essential. One is in place; the other is not" (Stiggins, 2002, p. 765). Instructional and program effectiveness (assessment for learning) can only really be measured at the group level: The larger the group, the more accurate the conclusions will tend to be. Group size can be increased by combining data from identical courses across a school, district, or state, or by including multiple years for an individual instructor's course.

Data analysis is one of the best tools available to inform and improve instruction. Too many educators have the attitude, "I taught it—they just didn't get it. The students just didn't put in enough effort." Truly effective teachers have the attitude, "I taught it—but they didn't all get it. I'll need to improve my approach next time." Performing item analysis of questions focused on essential learning objectives is one way to provide excellent feedback and critical direction for future instruction.

Data analysis can reveal topics that were difficult for students in general. Most often this would be accomplished by performing item analysis of individual questions on a common assessment. If the same topic weaknesses appear across many classrooms with a number of different instructors, it is likely that either the curriculum (texts and written materials) does not address those topics in sufficient depth or the instructors do not have strength in those topics. Data analysis could then lend direction to curriculum writing and staff development efforts. This assumes, of course, that the test items yielding those topics have been successfully screened for validity and reliability.

A Note on P-Values

The p-value of an item (item difficulty) is the probability that an average student will answer a particular question correctly. For a typical multiple-choice question with one correct answer and three distractors (and students with no background on the subject), the expected p-value is .25. This indicates that a student with no background has a 25% chance of correctly answering that item simply by guessing. However, if an item has a p-value of .25 after students have received instruction on the content addressed by the item, it would be considered an extremely difficult question. Analyzing the percentage of responses to each incorrect distractor in an item should also show an equal distribution of answers from the class, indicating that all distractors are equally plausible. Generally speaking, other than within safety tests, items with p-values above .90 should be removed from tests. Note that removing an item from a test does not indicate that the topic it covers is not important. It simply means that the item is too easy or difficult to provide much useful information about student learning. The majority of items on a properly constructed test, given to students who have had adequate instruction, would have p-values in the .40 to .70 range. (Readers can refer to Chapter 3 for a more in-depth discussion of item analysis.)

Instruction can also be improved through p-value analysis. If an item of known p-value shows a significantly higher score for a given class, the cause is very likely instructionally related. This provides direction for future instruction on that topic, or (in the case of formative assessments) for group remediation. Since the actual p-value will differ at least slightly for each group tested, p-value analysis can provide insights into the performance of different groups if it is run on each of a variety of subpopulations (ethnic groups, genders, etc.).

Measuring Course and Program Achievement of STL

Like most sets of comprehensive standards, the *Standards for Technological Literacy* cannot be adequately addressed within a single course. However, within a program, through exposure to a number of courses, the standards can be comprehensively addressed.

To assess the extent to which a program is delivering technological literacy, portions of the standards and a corresponding set of assessment items could be assigned to each course. Analysis of these data would pull results from standards-related items across all courses to determine how well the program addresses a set of standards. Additionally, since each student may choose a different route through a variety of courses, standards-related items could be linked to individual students across a variety of courses to determine how effectively a program of courses is delivering standards-based knowledge to individual students. Alternatively, a comprehensive set of standards-based items could be assembled into a summative assessment and delivered in a capstone course. Some method of assessing program achievement is essential in determining if comprehensive sets of standards or goals are being met; treating courses in isolation (which is too often the case) is not sufficient for evaluating programs.

ORGANIZING ASSESSMENTS: PLANNING FOR ANALYSIS

Analysis Implications Related to Data Gathering

The procedures used in gathering data have strong implications for the work needed later to perform analysis. First, an assessment plan should be developed that shows the number and type of assessments, the relationship of assessments to each other, and the formula for combining assessments into a holistic picture of student achievement. Data that is gathered can then be properly weighted and formatted at the outset. Additionally, beginning data gathering using a format that is compatible with analysis tools will save work in the end. For example, alpha characters—including pluses and minuses—are not friendly to databases and must be converted to numeric format if they are part of any electronic analysis. Entering data by writing scores in a grade book will require re-entry of that data into a database for efficient analysis, whereas initial electronic entry (directly into a computer or handheld PDA-type device) will save work in the end. Many electronic grade books have export functions that allow data to be easily moved into analysis tools. Scannable answer sheets work well for declarative testing, and a number of companies (for example, Scantron and NCS) produce data scanners and software that will automatically move scanned scores into an electronic grade book or school/district database.

Cleaning Data

Any time group data is analyzed, the data must first be "cleaned." This process usually involves creating a database or spreadsheet that contains all of the data elements or fields that will be used in the analysis. In large-scale assessments, this will involve thousands of students and each student record may have many associated fields or data elements. In large data sets, it is also likely to have students with common elements (i.e., the same names). Therefore, the first step in cleaning the data is to create a unique identifier for each student. The identifier could be a state or district student identification number, or a unique identifier created just for the analysis.

Once the unique identifier is attached to each student record, the next step is to search for and eliminate any duplicate records. Many software packages, such as Access, will have query or search features to find duplicate records. Duplicate records may occur for a variety of reasons. For example, students who were absent the day of a test may receive a score of "0" and then an actual score when they make up the test, thus creating two records for the same student.

Finally, the data will be analyzed by focusing on each of a variety of data elements (i.e., ethnicity, special education status, poverty status, pretest/post-test gain scores, and so on). As each element becomes the focus for analysis, one last data cleaning must occur within that field to eliminate any blank records existing in that element. In the case of comparing pretest and post-test score gains for large data sets, it is common to have many students who missed one test or the other, necessitating their removal from the data set. Failure to clean the data would create inaccurate statistics derived from any data analysis.

TECHNIQUES FOR DATA ANALYSIS

Analyzing Individual Student Data

This chapter's discussion thus far has assumed that each data element gathered can be converted to a point system that will be compatible with entry into a database or spreadsheet. However, a number of educational researchers prefer a holistic approach to scoring student work, and these techniques might be particularly adaptable to technology education.

Zemelman, Daniels, and Hyde (1998) assert that learning itself must be holistic. In describing the tendency of American educators to break ideas and concepts into smaller parts for instructional purposes, they state, "This part-to-whole approach undercuts motivation for learning because children don't perceive why they are doing the work. It also deprives children of an essential condition for learning—encountering material in its full, lifelike context. When the big picture is put off until later, later often never comes" (pp. 9–10). Instead, they argue that students will be most successful when learning is focused on the whole and sequenced from whole to part.

Similarly, data can be gathered in a holistic manner, scoring the "whole" in addition to, or rather than, the parts. Since technology education courses frequently employ lifelike contexts and often require a holistic product at the end, they easily lend themselves to this type of holistic scoring. At the least, a weighted scoring scheme could be employed to more heavily weight the holistic product. It could persuasively be argued that the scores on practice or partial work should not be factored in at all, since they might serve to penalize students who had achieved full success by the end of a course.

Measuring student growth in technology education courses is critical to improving instruction and programs. Unfortunately, long-term growth measurements have not been as easy to accomplish in technology education as in other curricular areas, such as mathematics and reading, which have curricula that span many grades (K–16). The vision of the ITEA, however, is that the STL become the next "core" curriculum area (ITEA, 2005, p. 28), and, like math and reading, that they span all educational grades. If that vision becomes reality, it will enable long-term growth measurements for technology education (and for technological literacy) as well.

Analyzing Student Group Data

There are many factors that contribute to the success or growth of individual students, a number of which are outside of the control of educators (health, sleep habits, work habits, family issues, etc.). However, when aggregating the results of many students, the greatest single contributor to success of a student group is the effectiveness of the instructor. It is well understood that not all teachers or teaching techniques are equally effective. Data analysis is the tool that can create a comparative base to identify which teachers or instructional methods are most successful. Essential components in this comparative process are as follows:
- The courses or groups being compared must be identical or must focus on the same outcomes/standards.
- The assessments and rubrics for each course/group must be the same.
- The data must first be cleaned, as outlined earlier in this chapter.
- Data summaries should be provided for the participants. The summaries must be disaggregated by the critical elements of the course outcomes/standards.

- The comparative process must be focused on improving instruction through data interpretation and collegial interaction. If the participants perceive the process to be focused on their own evaluation or on punitive ends, little gain is likely to occur.
- A process and time must be provided to facilitate discussions about the data.

Equal in importance to the process of analyzing data is identifying the focus of that analysis. Data analysis is often focused on key ideas that will either explain what has happened, predict future results, or guide change. Topics of focus might include change in results over time; growth trends or patterns; score or response distributions; gaps in performance/response between groups; interactions between groups or data elements (correlations); or achievement of standards.

Data Warehouses

Producing valid and reliable information through data analysis is a complex process. Generally, the simpler the process, the less reliable and valid the conclusions will be. To consider and gather all of the data elements that bear on a question or decision, a large set of data elements is required. For example, elements that correlate to test scores and student achievement include attendance, previous coursework and course grades, socioeconomic status, ethnicity, special education status, and age or grade. Few of these data elements are found in a typical teacher's grade book, or in any other single location. A data warehouse, in simple terms, is an electronic tool that retrieves the needed data elements from a variety of locations and assembles them in one location. Therefore, access to a data warehouse significantly increases the depth of analysis that can be performed. Additionally, a data warehouse could be assembled to allow aggregation of data across years, with multiple instructors and multiple institutions/schools. As with any statistical process, the larger the sample set or database, the stronger the conclusions tend to be.

Many commercial data warehouses are now available. However, for the typical technology education program, it may be easiest to simply export the desired data elements from whatever the institution's existing databases are and combine those data elements in tools such as FileMaker® Pro, Microsoft Access®, or Microsoft Excel®. Keep in mind that data privacy practices must

Data Analysis Techniques

be followed and that some data elements, such as socioeconomic status and student names, may not be accessible by all staff.

As an example of a commercial data warehouse, Anoka-Hennepin, a Minnesota school district of more than 40,000 students, uses the Switftknowledge data warehouse. This tool allows the district to assemble a wide variety of assessment scores, both from the current year as well as from past years, into one location, and to provide password-protected access to district staff. Since the scores from standardized tests are uploaded centrally, teachers can access that information without the need to enter the data themselves. Demographic and other data are also stored in this warehouse, making it possible for teachers and administrators to perform custom sorting and analysis to investigate their students' achievement by gender, ethnicity, special education status, or a host of other possibilities. Lists may then be printed out for use in creating groups for instructional intervention. In addition to individual classrooms, analysis may be focused on the grade level, course, building, or whole-district level, which provides information useful in making district-wide program and curriculum decisions.

A final advantage of data warehouses is the ability to quickly access data for the purpose of differentiated instruction. For example, if assessments are scored to provide data on specific critical topics or strands, and if the warehouse can be sorted on those strands, then the warehouse can be used to identify individuals or groups of students in need of focused topic review. Comprehensive data warehouses can even be integrally linked to instructional or curriculum elements so that the data can be used to first sort students into groups of common levels, and then link the groups to their areas of demonstrated need for remediation. McLeod (2005) states, "Schools that wish to fully realize the power of data-driven decision making, however, will move beyond the simple use of baseline data for goal setting and also will implement three other elements of data-driven instruction: frequent formative assessment, professional learning communities, and focused instructional interventions" (p. 2). McLeod also points out that formative assessment data typically do not reside in data warehouses, and suggests the use of an instructional management system, which may either run parallel to, or be integrated with, a data warehouse. Commercial management systems will often feature prebuilt assessments or item banks of questions, but these focus primarily on core topics such as math and reading, and a custom tool would need to be developed for technology education (ideally centered on the Standards for Technological Literacy). Some of the management system products also include handheld wireless devices to gather and transmit the data—a definite advantage in laboratory-based courses such as technology education.

Linking Data Elements

As mentioned previously, when working with group data, a unique identifier must first be assigned to each student, and then the data must be cleaned. The unique identifier must be present in each of the data sets to link the data elements. Once again, establishing a data warehouse will simplify this process. Elements of data to be linked should stem from the following four basic questions (Bernhardt, 2000, p. 5):
- Who are the students (demographics)?
- What do they know (student learning results)?
- What are they experiencing (school processes)?
- What do they perceive about the learning environment (perceptions)?

Perceptions are often obtained through anonymous surveys. However, if that data is to be used to its greatest extent, it must be linked to other elements through an identifier such as a student number, thereby sacrificing anonymity.

Applying Statistical Software

Most databases and spreadsheets include basic statistical functions (sums, averages, counts, simple charts, etc.). However, if more advanced statistical calculations are to be done with student data (quartile and decile calculations, standard deviations, correlations, etc.), it will be necessary to export data into a statistical tool such as SPSS or Minitab®.

Many institutions use scanning devices (such as the popular Scantron® system) to score multiple-choice and true/false type test items. These devices will produce basic statistics in addition to the scoring of the tests, but are generally limited in complexity. More advanced scanners will also produce statistics on item difficulty, and give counts for responses to each of the individual distractors in multiple-choice type questions. Some of the scanning/software products will integrate directly with electronic grade books and data warehouses, reducing the amount of data entry needed to maintain the system. A word of caution is in order, based on the author's experience: Scanners, like other technologies, must be maintained and cleaned regularly to function properly. The author has taken the same set of student answer sheets to different scanners and received different results, because dirty scanners or differences in sensitivity to stray marks produced inaccuracies. When student grades and program decisions are based on these data, efforts must be continual to guarantee accurate results.

A final, low-cost option is simply to employ the analysis/statistical tools within Microsoft Excel. Excel has features such as sorting, filtering, and counting as well as algebraic formula functions, charting, and Pivot Tables that will provide basic statistical tools with minimal training to virtually anyone. A variety of Web-based training programs are also available for Excel (such as the training offered by the University of Minnesota at www.schooldatatutorials.org). McLeod (2005) also suggests creating spreadsheet templates that are prepopulated with student names and have formula cells built in. These would simplify data input and analysis for staff within a building or district and could include an attached Pivot Table for displays of up-to-date data on demand without much technical knowledge. This is an especially viable solution for analysis of formative assessments.

Focusing on Technology Education: Data Analysis on STL, Federal, and State Mandates

Data analysis will ultimately only be valid if it focuses on the course and program goals. In turn, course and program goals must align with federal and state mandates (where applicable), and any technology education program should focus on the Standards for Technological Literacy. Therefore, data elements that reflect theses standards and mandates must be identified within the data sets analyzed. Since state and national standards are subject to periodic review and change, and data analysis for program improvement relies on stable measures over time, these concepts can be in conflict. The solution is to identify data elements that are likely to endure over time. Elements of design, for example, are not likely to change much as standards are revised. Data on big idea topics can then be followed over many years to monitor improvements in program effectiveness in those areas.

Disaggregating results to reveal—and reduce—achievement gaps between various student populations is an example of current mandates from federal and state levels for K–12 schools systems. Technology teacher education programs would also find it useful to focus data analysis on those elements, since political and community groups have identified reducing achievement gaps as a critical focus.

Reporting Data Analysis Results

Results of data analysis need to be shared with a variety of audiences, each of which has a different need for the data and different levels of expertise in data interpretation. Different data representations should be created for each of the following audiences and purposes:
- Students and parents need data to document the educational accomplishments of students.
- Teachers need data to analyze and improve instruction.
- Departments need data to market their program and identify needs.
- Institutions need data to verify the viability and success of programs.

Each of these audiences will require a different type of data presentation. Common to all audiences, however, is that they will only want the data that pertains to them, and that data must be presented and communicated effectively in the simplest way. Graphs and charts tend to communicate more effectively to a broad audience than do tables or lists of data; however, large quantities of data are often most efficiently displayed in a table.

Communicating Data Analysis Results to Students and Parents

When communicating with parents it is essential to provide answers to their typical questions, such as: "What has my son/daughter learned compared to a goal or standard?" and "How is he/she doing compared to other students?" The latter question typically requires analysis that establishes local or national norms and then expresses student progress either in terms of percentile ranking or comparison to a group average. If courses are sequential in nature and address the same learning construct, then student growth can also be reported in comparison to local or national norm groups.

The format of parent and student reports is an important consideration if communication about student progress is to be effective. Critical features of effective reports include the following:
- student identifiers (name, birth date, student number, gender)
- course identifiers
- limited text (especially with limited English proficient students/families)
- clearly labeled data
- charts or graphs, if appropriate
- keys or definitions of terms
- dates of assessments
- student status, if appropriate (e.g., pass/fail, prerequisite met, etc.)
- contact information, in case further information is sought
- review of format by a parent group and revision before widespread use

Report Formats for Instructional Improvement

Two questions need to be answered if data analysis is to be used to improve instruction: "How well did a group of students do compared to course goals or standards?" and "How much change in learning took place as a result of this course?" Answering the first question requires that assessments and analysis identify and extract data on specific criteria (the critical attributes/concepts being taught, usually referred to as topics or strands within an assessment), and that they describe student status compared to a fixed goal. Answering the second question requires a pretest/posttest type of analysis focused on the critical attributes.

A final key to improving instruction through data analysis lies in the belief that, ultimately, if learning has not occurred to the extent desired, it is the job of the instructor to change the situation (as opposed to the common belief that "the students just didn't get it" or "students need to put more work into this class"). In pursuing needed changes, instructors must formulate a number of questions regarding their instruction. "Was I equally effective in teaching all groups of students?" "Was I successful in enabling students to learn the most critical aspects of this course?" Assembling data into a sortable electronic format (such as Microsoft Excel) is a very effective way to answer these questions. Each question can be explored through data sorts in a relatively short time, enabling instructors to explore these questions (and answers) on their own.

Report Formats for Building and Program Summaries

Where electronic spreadsheet formats work well for individual teachers to analyze data, in most cases, hard-copy paper reports are better suited for building and program summaries where the intended audience includes many people representing a broad cross section of the community (many of whom will have relatively little experience in interpreting data). Rather than providing an audience with an ocean of data, it would be more effective to first identify the questions to be answered or messages to be communicated, and then create paper reports that effectively build the case in answering those questions. The questions may focus on accountability concerns, such as: "Are students in this institution achieving the desired results?" and "What is the extent of the achievement gap between student groups at this school?" The questions could also focus on accumulating evidence that a need exists, such as: "How does funding for this institution compare to similar institutions?" and "Is the laboratory equipment in this program preparing students for the world they will enter as graduates?"

At the building or program level, when there is a broad audience, Holly (2003) recommends creating reports that can stand alone and
- be relatively brief.
- use appropriate analysis of data, without lengthy descriptions.
- focus on important data rather than data that is simply "nice to know."
- present data that can be understood by the average person.
- show data in charts, graphs, and other pictorial representations.
- provide succinct analyses of data that offer interpretations of findings.

A report that incorporates these suggestions would identify the purpose of the report (in other words, the question to be answered); would limit the data to focus on that purpose; would display the data in an easily understood format; and would include just enough text to clarify the findings of the data.

CONCLUSION

Data analysis is an absolutely critical tool in educational improvement. While analysis in itself will not solve today's educational challenges, it is probably the only tool that will tell us what really does work. Every educator and every educational system needs to engage in data analysis, and this chapter suggests a number of ways to accomplish this task.

Analyzing assessments for validity and reliability is an essential first step and must be performed on both declarative and procedural assessments. Suggestions for accomplishing this are woven throughout this and other chapters.

Data at the individual student level is available to all teachers, and includes information from both formative and summative assessments. Analysis at this level is most useful in measuring individual student growth (improvement over time) and student progress toward fixed goals, such as standards. The results of this analysis find their way into reports and into feedback to both students and parents. They also provide excellent insight into the individual needs of students.

Data at the group level adds a new dimension and allows teachers to critique their own effectiveness and make instructional decisions on how to better approach their course content. At the program and district level, group analysis reveals the effectiveness of programs, differences between teachers, and sheds light on staff development needs as well as progress toward broad goals such as developing a technologically literate student population. Group data analysis, however, introduces a host of data treatment considerations, and this chapter addressed the need for "analysis-friendly clean data." Data warehouses and other solutions are suggested for managing large data sets.

In the end, analysis of data from educational assessments needs to be effectively communicated to the appropriate stakeholders and customized for the unique needs of the technology education classroom.

DISCUSSION QUESTIONS

1. What steps could be taken to improve the reliability and validity of data being gathered by a classroom teacher? By a district or state?
2. What steps can be taken to prepare data for efficient and valid analysis?
3. Among the different purposes of analyzing data, when would individual student data be used and when would group data be preferred?
4. What are the various purposes of performing classroom, district, state, and national level analysis?
5. In what ways are technology education courses unique compared to courses in other departments, from a data analysis perspective?

REFERENCES

Bernhardt, V. L. (2000). *Designing and using databases for school improvement.* Larchmont, NY: Eye on Education Inc.

Helberg, C. (1995). Pitfalls of data analysis. Paper presented at the Third International Applied Statistics Conference, Dallas, TX.

Holly, P. J. (2003). *Creating a data-driven system.* Princeton, NJ: Educational Testing Service.

International Technology Education Association (2000). *Standards for technological literacy: Content for the study of technology.* Reston, VA: International Technology Education Association.

——— (2005). *Technological literacy for all: A rationale and structure for the study of technology.* Reston, VA: International Technology Education Association.

McLeod, S. J. (2005). Technology tools for data-driven teachers. White paper. http://www.scottmcleod.net/dddm_resources/ (accessed February 20, 2006).

Stiggins, R. J. (2002). Assessment crises: The absence of assessment for learning. *Phi Delta Kappan* 83:758–765.

Stiggins, R. J., J. A. Arter, J. Chappuis, and S. Chappuis (2004). *Classroom assessment for student learning.* Portland, OR: Assessment Training Institute.

Wiggins, G. P., and J. McTighe (2005). *Understanding by design.* 2nd ed. Alexandria, VA: Association for Supervision and Curriculum Development.

Zemelman, S., H. Daniels, and A. Hyde (1998). *Best practice: New standards for teaching and learning in America's schools.* 2nd ed. Portsmouth, NJ: Heinemann.

Applying Classroom Assessment Data: Communicating Results and Modifying Instruction

Chapter 7

Jared V. Berrett
Steven L. Shumway
Brigham Young University

CHAPTER OVERVIEW

The purposes of this chapter are to describe various methods of communicating assessment results to the primary participants of the learning environment (i.e., students, parents, administrators, and the community) and to discuss how assessment results might effectively be used to modify classroom instruction. In support of the first purpose, strategies are provided that will help teachers communicate both formative and summative assessment results to students, parents, and administrators. Additionally, a reflective practitioner model will be presented as an effective method for teachers to use assessment data to improve classroom instruction. It is our hope as authors that teachers might think of assessment as an iterative process, one which constantly requires them to reconsider and modify their instructional practice. This process begins with recognizing a need and having a desire to become more reflective and sensitive to the overall instructional requirements of students; being willing to modify instruction; and, finally, taking action and making changes to instruction based on the data and information generated.

COMMUNICATING ASSESSMENT RESULTS

Communicating Through Formative Assessments

Although assessments that are summative in nature (e.g., standardized tests, exams, portfolios, and other assessment tools that provide information relative to student understanding, following a sustained period of instruction) have been a natural focus within the educational profession,

Applying Classroom Assessment Data: Communicating Results and Modifying Instruction

there are other assessments that teachers conduct which are equally important. Known as formative assessments, these assessments are typically categorized as repetitive and ongoing (e.g., observations, interviews, class discussions, and so on) and are conducted throughout the instructional process, with the main purpose of providing information and feedback to the students and teachers, relative to small segments of course material. These assessments are important to the day-to-day educational process but are often overlooked by classroom teachers. It is essential to consider how teachers communicate the results of these assessments to students and others.

Classroom Vignette:

The students in Ms. Wagner's class are excited. They are almost ready to begin testing the bridges they have made as part of a construction unit. As the students finish some last-minute paper work, Ms. Wagner checks to see that all are making preparations to test their bridges. She looks around the room and sees that while most of the students have almost completed their assignment, a few are still struggling. Alex is having difficulty with his paperwork—he seems to always struggle with reading and writing. Three other students are just now gluing their bridges. They will not be ready to test today. Amy and Jill have finished everything and are bothering some of the students at their table. From her vantage point, Ms. Wagner can see at least two students with design-rule violations—didn't they read the rules? Why have only three students decided to use abutments? She realizes that most of the students must not have understood that part of the discussion.

Ms. Wagner comments to Amy and Jill, "Would the two of you be willing to come and set up the testing device while I help some of the other students?" Amy and Jill look pleased that Ms. Wagner thinks they are capable of this extra assignment and hurry over to help. Ms. Wagner moves toward Alex's desk but realizes that if she continually hovers around the students that struggle, they will think that she doubts their ability to complete the assignment. Pausing slightly, she stops at Jim's desk to see how he is doing. She knows that Jim always does excellent work and probably does not need help, but Jim sits next to Alex, and this will allow her to move to help Alex next. Ms. Wagner is tempted to tell Alex that she sees several weaknesses with his bridge, but instead she asks him several questions about his design so that he will perform a self-assessment of where he thinks his bridge will finally break. During this brief exchange, it becomes evident that Alex has questions about his design and how to complete the paperwork. Ms. Wagner provides feedback by modeling the process and then reminds him of some methods they had discussed for strengthening a bridge. After helping Alex, she moves throughout the room in an "almost" random pattern. Ms. Wagner makes sure she observes all students, paying particular attention to, and helping, those needing extra assistance.

While Ms. Wagner is in the process of facilitating her classroom processes as exemplified in the vignette above, she does two things: provides formal feedback in direct response to questions and provides students with informal cues through mannerisms, body language, and questions. These informal communications are very powerful methods of communicating her assessments back to her students. They impact both the conscious and subconscious attributions students make for their successes and failures, a process well documented in education literature (Rosenthal, 1991; Good and Brophy, 1987; Weiner, 1986). However, teachers must be careful that they don't inadvertently communicate damaging messages to their students through these informal formative assessments or through interactions that happen daily in the classroom.

For example, if a teacher consistently asks easy questions of certain students in the class, or if a teacher repeatedly hands out assignments to the class and then moves automatically to the students she knows will need help, the message that might be sent to these students is that the teacher doubts their abilities. To avoid this unintended communication, teachers might employ a selective randomization process when helping students in the lab.

Teachers must also take care when expressing sympathy or excessive praise toward any given student. When a teacher expresses sympathy toward a student who has performed poorly, the student might equate sympathy with low ability. In addition, excessive praise, particularly for success on relatively easy tasks, can also be equated by students to mean that their abilities are lacking (Weiner, 1986). Instead of sympathy or excessive praise toward the individual, providing students with focused feedback relative to their work will help. In particular, using positive feedback in which the teacher provides comments about student learning relative to the subject (rather than to the students themselves) communicates a message that the teacher assesses the students' abilities and performances to be good (Weiner, 1986).

Communicating Through Self-Assessment

In December 1998, the National Research Council (NRC) released *How People Learn*, a report that synthesizes research on human learning. One of the key findings in this report is that "a metacognitive approach to instruction can help students take control of their own learning by defining learning goals and monitoring their progress in achieving them" (NRC, 2000, p. 13). Related to this finding is the concept that teachers might effectively communicate the results of assessment by teaching strategies and by providing opportunities for their students to "monitor their own progress." In this approach, students would monitor their own understanding, make note when additional information is needed, and determine if new information is consistent with what they already know (International Technology Education Association [ITEA], 2004).

Many strategies have been developed by educators to help their students learn metacognitive methods of assessing their own learning. Technology educators can familiarize themselves with these strategies and determine how each might be used in a technology education setting. While it is beyond the scope of this chapter to provide a detailed description of each, Buehl (2005) has identified and explained many of these strategies, which include Learning Logs, Three-Minute Pause, Think/Pair/Share, Reflect/Reflect/Reflect, and Self-Monitoring Approach to Reading and Thinking (SMART).

Many teachers do not incorporate self-assessments into their assessment strategies because they feel their students are not capable of making accurate assessments. Other teachers who use student self-assessments may be unsuccessful because they make the mistake of having students simply "grade" themselves without having a formal process or criteria for assessing their understanding. Students must be coached on what goes into an assessment before they are set free to grade themselves. For instance, is effort part of the assessment? What about competency? Which is more important: creative solutions or following directions? Many students are uncomfortable with the thought of self-assessment because they are rarely given the opportunity. Our experience shows that self-assessment can be successful if students are given a framework for how they might assess their own work.

For example, if Ms. Wagner wanted to formally involve her students in an opportunity for self-assessment during the construction unit, she could (a) provide a rubric for her students that contains items for assessment, with accompanying performance criteria and descriptors; (b) teach the class how to correctly use the rubric by modeling the assessment process for the students; (c) have students conduct a self-assessment of their work and write in their score; (d) assess the student work herself, using the same rubric; and (e) hold a short conference with individual students to discuss discrepancies if there are large differences between the teacher and student scores. The power in this process is that both the teacher and the students are able to reflect on student progress relative to the learning objectives. Although this process is time consuming, the lifelong learning attributes gained may outweigh the immediate pressures of classroom management.

Communicating Through Summative Assessments

With a renewed emphasis on accountability and high-stakes testing, strategies by which teachers effectively communicate the data from these summative assessments are an important concept to consider. Summative assessments, especially when they combine data from a variety of measures, allow teachers, parents, and other school officials to assess student understanding relative to the content area. One of the challenges thus becomes how you can effectively communicate the results of these assessments to the various educational stakeholders.

Communicating to Students

Communicating assessment results at the completion of a project, assignment, or curricular unit is often viewed as an "end-all" activity for both teachers and students. Unfortunately, it is sometimes assumed that the score or grade that summarizes the student performance is the only communication that matters. In actuality, the feedback that should accompany the score is probably more important when considering student learning. In a meta-analysis on student learning, Walberg (1991) found that when teachers take the time to give students corrective feedback on their work, student learning increases significantly (Effect Size=0.94). Feedback should be specific and accurate. General feedback like "Nice robot, Billy" can make a student feel good about his work but doesn't necessarily help him improve. Instead, teachers must provide specificity, such as: "The position of the fulcrum for the elbow on your robot provides excellent reach."

When communicating assessment data to students, it is not only important to provide accurate and specific feedback, but the assessments themselves should be fair and equitable. Teachers should assess student learning against specific criteria, make expectations clear to the students, and have multiple assessments serve as indicators for technological literacy (ITEA, 2004). Having a strategy for conducting assessments, such as use of a rubric, communicates to the students what is expected for a particular assignment and decreases the opportunity for subjective grading.

In addition, assessment scores should be easy for the students to interpret and understand. Some teachers violate this concept by having a different point system for various activities with a complicated weighting system for determining grades. The teacher may communicate the assessment results, but students are often mystified as to what it means. If possible, assignments should be made of equal value (e.g., 10, 25, or 100 points), or a percentage system that is easily understood should be used. Finally, teachers need to take the time to teach their grading system to the students to help them understand what the scores mean. This allows students to make educated decisions about their performance and about the consequences low or high performance will have on their grades.

Another suggestion is to communicate assessment data from a productive rather than a punitive approach. In this approach, instead of deducting points for incorrect items, you reward points to students for items they have completed correctly. This communicates to the student that they are being evaluated on what they have learned, rather than on what they have not learned (Smith, Smith, and De Lisi, 2001).

Finally, when handing back scores, papers, or assignments that have sensitive information on them, teachers should use student numbers or have students put their name on the last page of a document. This technique can also be helpful to the teacher to avoid biases when grading student assignments. Regardless of the method used, it is essential to keep student records confidential, and teachers should work with the school administration to determine the best way.

Communicating to Parents, School Officials, and the Community

Technology tools such as e-mail, school Web sites, conference phone calls, and electronic grade books are all viable ways to communicate assessment data to parents and other school officials. These technologies present a potential trade-off: Parents and others will feel better informed and more involved in the student's education; however, teachers may feel that these tools require additional time or that unrealistic expectations regarding the quantity and quality of communication have been set.

A very traditional approach to communicating with parents is through parent-teacher conferences. These conferences can take place through a variety of methods, such as e-mail, letters, phone calls to parents, or the official parent-teacher conferences held periodically at the school. Regardless of the forum, there are a few rules that teachers should follow to best communicate assessment results during conferences. First, be prepared. No matter what you are doing in your classes, keeping good records and documentation of student work is essential. These records do not have to be generated by the teacher, but they do need to be accessible.

Second, know your students' names and keep an up-to-date summary of each student's work. Showing the parents a student portfolio or having student work on display is an excellent way to communicate to the parents how their child is doing in the class. If your school has an Internet-based system for reporting student progress, learn how to use it and keep scores up-to-date. Many parents are beginning to rely heavily on the Internet to track their children's progress, and they are frustrated by teachers who refuse to use the system properly.

Next, make assessments easy for the parents to understand. At the beginning of the year, send a disclosure statement, in which the grading system is explained, home with the student. If you are going to be using a

Applying Classroom Assessment Data: Communicating Results and Modifying Instruction

rubric to assess student work, make a copy of this rubric available to the parents as well as to the student in order to communicate expectations. Finally, if data from a standardized (normalized) test is sent to the parents, some type of key that helps the parents understand the data should be provided, or a conference with the parents should be arranged.

Show the parents that you care about the success of their child. During a conference, be able to first share some of the child's strengths, and then be prepared to discuss improvements the child needs to make. Focus on behavior and facts (have examples of the child's work or anecdotal notes to share as examples) instead of making broad generalizations. Ask the parents what their concerns are. The parents might be able to tell you things that will help you in your role as the teacher. This might also enable you to invite the parents to be more actively involved in the student's education. Finally, be concise when communicating assessment results. This is especially true during parent-teacher conferences. If additional time is needed, set up another conference.

An effective way technology educators can communicate to school officials and the community at large regarding student accomplishments is to develop a public relations agenda. One of the best ways to do this is to allow these individuals to observe students actively engaged in the learning process. Some examples include

- forming a Technology Student Association (TSA) club.
- having students interview an important adult in their life about the development of technology and how different technologies have changed their daily routine.
- involving your technology education class in a community environmental awareness project (e.g., cleaning up a small stream or cellphone battery recycling).
- setting up a public display of student artifacts.
- inviting specialists from the community to do collaborative activities with your students. See that the local newspaper is informed.
- inviting school officials and persons from the community to your classroom to judge competitions or witness student presentations.

The opportunities to communicate to school officials and the general public what your students are learning are many and varied. None of this can happen unless the teacher takes the initiative to do some extra planning and incorporates the ideas into the curriculum.

MODIFYING INSTRUCTION

Becoming a Reflective Practitioner

One of the underlying purposes of assessment is to improve instruction (Smith, Smith, and De Lisi, 2001). Suggestions on how teachers can become reflective practitioners and use the data gathered from various assessments in order to make improvements to classroom instruction is the focus of this section.

One strategy for becoming a reflective practitioner is to develop a personal professional portfolio as described by Reineke (1998). The purpose of the portfolio is to help the teacher record, monitor, and refine instructional strategies and assessment procedures. Engaging in making a personal portfolio does involve some personal risk, since the teacher will be collecting data related to instruction and assessment practices from both students and colleagues, and not all this data will be positive. The major components of the portfolio are shown in Figure 1 and are described below.

Figure 1: Components of a personal professional portfolio.

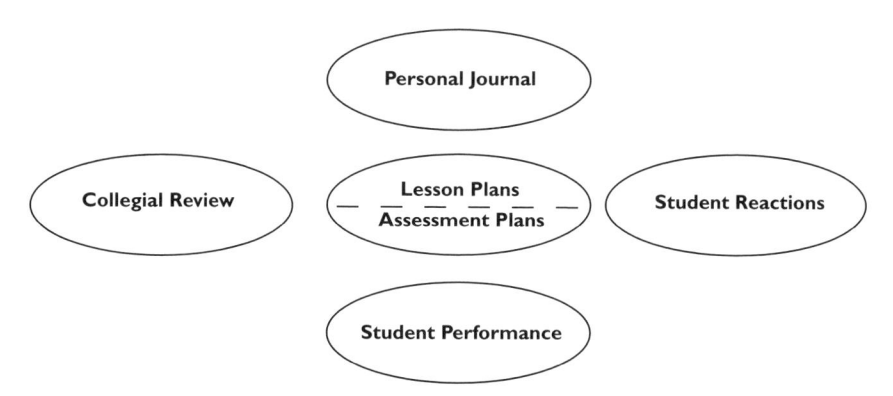

In the collegial review, have colleagues observe your classroom to provide data related to the instruction and assessment climate in your classroom. Have them help you determine if what is being taught and what is being assessed are congruent. Are students willing to ask the teacher questions if they do not understand material? What are the students' reactions to instruction and assessment? Additionally, student performance and student reactions should be included. Both formative and summative evaluations of student work should be represented. Your observations, as well as observations of colleagues related to how students have reacted to instruction and assessment, are essential to the portfolio. Do students readily participate in class? What are some of the casual comments that students have made regarding instruction and assessment? Are the students engaged during instruction? What are some concepts that students have understood well? What are some they have not understood well?

The purpose of the personal journal is to provide the teacher with an opportunity to reflect on instruction and assessment practices, including information collected from collegial reviews, student performances, and student reactions. Teachers should focus initially on collecting descriptive, rather than judgmental, data for reflective purposes. The data that has been collected in the form of the personal portfolio can then be utilized to make systematic formal changes to lesson plans and assessment strategies. In fact, perhaps the most important concept is that by developing the portfolio, reflection and systematic change become a formal process, rather than the informal process many teachers use.

Chown and Last (1993) suggest that teachers should not only recognize how they and others—including their students—think about their classroom experience, but that teachers should take action based upon that reflection. By identifying areas of need and potential sources of new ideas and understanding to address those needs, a course of action should become evident. Morin (2004) has identified other strategies for reflective practice.

- Start with standards: Review your state standards or the *Standards for Technological Literacy* periodically to see if your instructional material, lesson plans, unit plans, or course outlines reflect the necessary content.
- Audio or videotape yourself teaching, and analyze the recordings. Consider the positive over the negative, and build upon positive experiences.
- Once a week, select a student you want to help. Flip through the student's file to get to know him or her better.
- Rate the relevance of your course content. Consider one or two elements of your curriculum, how you assess these elements, and challenge yourself to find ten reasons for teaching students this content.

Being a reflective practitioner begins with solid instructional strategies. Determining what students are supposed to be able to do or know at the end of the course, prior to thinking about assessment, is imperative. How will we know what it is we want to improve if we do not know what it is we are trying to accomplish? Setting the goals for a course might vary widely according to state objectives, personal interests, and community needs. These goals may focus on knowledge we hope the students will gain, habits of mind we hope they will adopt, or skills we hope they will develop—but irrespective of the goal, good instructional design answers the question of what to assess.

A second question that is imperative to reflective practice is, "Who are we teaching, and how do they learn?" More and more, teachers are facing diverse student populations, making it very difficult to close the student achievement gap in the classroom. By the time you consider how a student learns and what learning styles different students might favor, the first steps of assessment can become pretty complex. Kohn (2004) suggests that while every teacher possesses a theory of learning, there often exist profound differences between what people espouse as excellent learning and teaching practices and what strategies or practices are used to assess knowledge. For example, believing that students learn best through inquiry and problem solving, but only providing rote tests and fill-in-the-blank worksheets for assessment purposes because they are easy to grade, creates a tension between the belief and practice that must be resolved. In other words, good teaching goes hand in hand with good assessment.

CONCLUSION

There are many effective strategies for communicating both formative and summative assessments to students, parents, school officials, and even the general public. In order to improve the teaching and learning process, technology teachers need to become familiar with these strategies and formalize a process for their successful implementation in the classroom. In addition, teachers need to understand that effective assessment and good instruction are codependent. By becoming reflective practitioners, teachers can formalize the process by which they use assessments from various sources to make systematic improvements to classroom instruction.

Applying Classroom Assessment Data: Communicating Results and Modifying Instruction

DISCUSSION QUESTIONS

1. What are some of the opportunities teachers must consider in their instructional practices to improve summative assessments?
2. What are some of the ways teachers communicate unintended messages to their students as they perform formative assessments of student learning?
3. From your personal experience, are middle school or high school students mature enough to engage in the self-assessment process? Why or why not?
4. What are some of the reasons why teachers are hesitant to incorporate formal reflective practices, such as developing a personal professional portfolio?
5. What are some of the strategies that you have found to be successful when communicating assessments of student learning to parents?
6. How are technology educators uniquely positioned, with their hands-on instructional content, to make meaningful connections to student learning and curriculum through well-crafted assessment strategies?

REFERENCES

Buehl, D. (2005). *Classroom strategies for interactive learning.* Newark, DE: International Reading Association.

Chown, A., and J. Last (1993). Can the NCVQ model be used for teacher training? *Journal for Further and Higher Education* 17(2): 15–26.

Good, T. L., and J. E. Brophy (1987). Motivation. In *Looking in classrooms,* 4th ed., edited by T. Good and J. Brophy, 173–215. New York: Harper & Row.

International Technology Education Association. (2004). *Measuring progress: A guide to assessing students for technological literacy.* Reston, VA: International Technology Education Association.

Kohn, A. (2004). *What does it mean to be well educated? And more essays on standards, grading, and other follies.* Boston: Beacon Press.

Morin, J. (2004). 20 ways to reflect upon your practice. *Intervention in School and Clinic* 40(2): 111–15.

National Research Council. (2000). *How people learn: Brain, mind, experience, and school,* edited by J. D. Bransford, A. L. Brown, and R. R. Cocking. Washington, DC: National Academies Press.

Reineke, R. (1998). *Challenging the mind, touching the heart: Best assessment practices.* Thousand Oaks, CA: Corwin Press.

Rosenthal, R. (1991). Teacher expectancy effects: A brief update 25 years after the Pygmalion experiment. *Journal of Research in Education* 1:3–12.

Smith, J., J. Smith, and R. De Lisi (2001). *Natural classroom assessment: Designing seamless instruction and assessment.* Thousand Oaks, CA: Corwin Press.

Walberg, H. J. (1991). Productive teaching and instruction: Assessing the knowledge base. In Effective teaching: *Current research,* edited by H. C. Waxman and H. J. Walberg, 33–62. Berkeley, CA: McCutchan Publishing.

Weiner, B. (1986). *An attributional theory of motivation and emotion.* New York: Springer-Verlag.

Aligning Curriculum and Assessment in the Development of a Statewide Accountability System

Chapter 8

Rhonda Welfare
Thomas Shown
North Carolina Department of Public Instruction

OVERVIEW OF CAREER-TECHNICAL EDUCATION IN NORTH CAROLINA

For more than a decade, North Carolina has focused on developing an assessment system that closely links curriculum and assessment in career and technical education (CTE) courses. Technology education students in North Carolina begin the school year with a clear list of the material that will be covered during the course and knowledge of the objectives to be assessed. Results from these assessments provide a measure of student achievement that becomes part of the state's accountability process, and they can be used as part of instructional improvement efforts.

Development of North Carolina VoCATS

Thirty years ago, North Carolina began a curriculum initiative that would eventually become the Vocational Competency Achievement Tracking System (VoCATS), the umbrella under which CTE curriculum is developed, disseminated, and implemented. As it was originally conceived, North Carolina's effort focused on developing lists of outcome behaviors expected upon completion of vocational programs. Some teachers maintained paper records and manually tracked student performance by placing a check mark in the appropriate space on a chart when students demonstrated the desired outcome.

When used, this was an effective, but not very efficient, use of teachers' time. Many teachers found the system too cumbersome to be helpful. Even for those teachers who did track student progress, at the conclusion of the year, the paper charts were thrown away to make way for new charts, so no permanent record of student proficiency was created.

Aligning Curriculum and Assessment in the Development of a Statewide Accountability System

By the late 1980s, the system had begun to evolve into a broader effort to standardize curricula statewide. Using federal Program Improvement funds, North Carolina began to develop what were called "blueprints," or lists of specific outcomes and objectives—of selected programs—to be implemented in pilot school systems. The blueprints, created by teams of teachers with input from business and industry representatives, also indicated additional information about each objective. This additional information also included each objective's relative importance in the overall course, its classification using Bloom's Taxonomy (Bloom, Krathwohl, and Masia, 1984), and its general relationships to other subject areas.

The blueprints provided a framework for instruction. Curriculum guides and expanded course outlines, which North Carolina had long provided to assist teachers, were revised to align with the course blueprints. The blueprints also made it possible for administrators to analyze areas in which teachers needed additional training and which facilities, equipment, and other resources were inadequate. Test-item banks were developed and aligned with the blueprints again to provide an additional tool for teachers to use to analyze student progress. Eventually, multiple-choice tests were created and provided to school systems upon request. Local administrators used these tests to understand how their students were faring on the blueprint objectives. Data were not collected at the state level nor were any state reports generated.

From the initial small set of courses that were included in VoCATS, the system expanded to include more courses and more school systems. As the system grew, a technology-based solution was required to make managing data possible. In 1991, the state requested proposals from instructional management software vendors to recommend how to proceed. There were three specific requirements for the software:

1. The software needed to use state-generated, test-item banks.
2. Teachers had to be able to add items to the banks.
3. The software had to be able to display and print graphic images.

After considerable review, CTB/McGraw-Hill's TestMate family of products was selected. This software included Curriculum Builder, to develop and manage test-item banks; TestMate, to scan and score tests and generate reports; and TestTracker, to monitor student performance across multiple tests. The software was written for DOS and ran on a PC platform only.

VoCATS curriculum materials were used widely in the state, but many teachers still did not focus on the state frameworks. Local administrators were looking for ways to encourage teachers to accept VoCATS and use the state-developed curriculum. North Carolina CTE staff members carried out required program monitoring, but monitoring was limited to what could be done in meetings and by observation. No formal assessment was made of student performance. Relatively few programs could be visited each year, so the impact of program monitoring was limited.

VOCATS ASSESSMENT COMPONENT EXPANSION

By 1992, tests were provided for nearly all CTE courses in North Carolina, including technology education. Local administrators began to request that the state collect data and produce a statewide report to provide a frame of reference for local data. The state staff observed wide variations in how VoCATS was being implemented and wanted a better understanding of the process. The Carl D. Perkins Vocational and Technical Education Act of 1990 provided a means to satisfy those needs. For the first time, this law, which was tied directly to receipt of federal vocational funds, required local school systems to document student proficiency in technical competencies.

North Carolina negotiated a plan with the U.S. Department of Education that required local school systems to administer pre-CTE and post-CTE tests to all CTE students and to report local results for mastery and gain. Mastery was defined as the percentage of students who scored at or above 80% correct on the posttest. The gain target was defined as the percentage of students who moved at least 60% of the distance between their pretest score and their total possible posttest score of 100% correct for introductory courses (called Level I courses). For Level II courses, the gain target was defined as the percentage of students who moved at least 40% of the distance between their pretest score and 100%. To ensure equal access, the state purchased one set of the TestMate family of software for each school district.

Implementation of the accountability system encountered serious obstacles. Although data collection and reporting were done using computers, the software was quite cumbersome and difficult to use. Reports were generated in hard copy only and could not be retrieved as data files to be further analyzed or aggregated and analyzed at a different level. As a consequence, no statewide reports were available. Local reports were kept on file at the state, but there was no way to monitor how they were used to improve instruction. Perhaps most seriously, teachers learned to maximize their students' gain scores by encouraging them to answer on the pretest only those questions about which they were sure, but on the posttest to answer all questions, even if their answer was only a guess.

When the Perkins law was reauthorized in 1998, North Carolina was prepared to make several rather substantial changes to the accountability system. The state purchased an additional piece of CTE software, TestMate Clarity®, for each school system. This software allowed the processing of reports in a Windows environment, and it was much more user-friendly than the earlier DOS software. North Carolina also established two Web sites, both of which are accessed via secure login. One site allows data to be uploaded to the system, and the other is for posting of local plans and reports based on the uploaded data.

North Carolina's system of tracking student technical attainment was modified to more closely resemble the state's system for reporting student performance in traditional academic areas. The pretest was dropped. Student posttest scores for every CTE course statewide were combined, and a scale was developed that identified student performance as Level I (44% or lower correct), Level II (45–64% correct), Level III (65–81% correct), or Level IV (82% or higher correct). Baseline data were collected in 1998–1999, and a statewide target was set that by 2002–2003, 58.2% of the students enrolled in CTE courses would reach Level III or Level IV proficiency on the post-assessment. (That goal has since been extended. For the 2005–2006 year, the target was for 60.5% of the students enrolled in CTE courses to reach Level III or Level IV proficiency.)

Data from the system permits local CTE administrators to pinpoint areas of greatest need and direct resources to those needs. Results can also be disaggregated by special populations categories to be sure all groups of students are being served. Special populations students include those who are handicapped (Exceptional Children), limited English proficient, or who have been identified as at risk due to academic or economic disadvantages. Locally, data are generated that indicate student performance on specific objectives by teacher or even by class. These data can be compared to state results by objective, and teachers can select areas where their student performance is below the average for students across North Carolina. Teachers can develop classroom strategies to improve student learning relative to these objectives, and administrators can target funds for equipment and professional development directly tied to student performance.

In addition to technical attainment data, North Carolina develops reports showing academic attainment by program area. Graduates who meet the state definition of "concentrator"—having earned at least four credits in a career pathway, with at least one of those credits in a second-level course—are assessed. These students take four of the subtests of the ACCUPLACER or ASSET tests, two nationally recognized examinations used primarily for community college placement. The four subtests are reading, writing, numerical skills, and elementary algebra. Results of the subtests are then analyzed to determine what percentage of test takers scored at or above the national average.

Targets are set for student performance within each subtest. For example, for the 2002–2003 year, the goal was for 51.3% of the test takers to score at or above the national average on the reading subtest. (That goal has since been extended. For the 2005–2006 year, the target was for 56.8% of the test takers to score at or above the national average on the reading subtest.)

HOW THE VOCATS SYSTEM WORKS: A CASE STUDY

Overview of Technology Education in North Carolina

In North Carolina, technology education is part of career and technical education. The 11 technology education courses (Table 1) are listed within the current CTE *Standard Course of Study Guide* (2002). There is only one course serving the middle grades (seventh and eighth)—Exploring Technology Systems. The 10 others are high school courses, with Fundamentals of Technology serving as the gateway to the four systems courses. Technology Advanced Studies serves as the culminating course. Since the release of the current *Standard Course of Study* in 2002, Project Lead the Way courses have been added to the technology education scope and sequence.

Aligning Curriculum and Assessment in the Development of a Statewide Accountability System

Table 1. Scope and sequence, with statewide enrollment data from 2004–2005.

Middle School	High School			
Grades 7–8	Level I	Level II	Level III	Level IV
Exploring Technology Systems (57,092)	Fundamentals of Technology (12,665)	Communication Systems (2,153)	Principles of Technology II (491)	Technology Advanced Studies (1,268)
		Manufacturing Systems (753)	Scientific and Technical Visualization II (383)	
		Structural Systems (1,070)		
		Transportation Systems		
		Principles of Technology I (2,377)		
		Scientific and Technical Visualization I (731)		

During the 2004–2005 school year, approximately 57,092 students were enrolled in the middle grades course and approximately 22,630 in all high school courses combined, totaling 79,722 (excluding approximately 1,000 enrolled in Project Lead the Way courses). These roughly 80,000 technology education students are slightly more than 9% of all students served by CTE.

The primary focus of the North Carolina technology education program has been to develop general technological literacy and, for the most part, is distinct from the trade and industrial education programs, which focus on specific skill development areas such as drafting, cabinetmaking, automotive technology, and other trades and occupations. The exceptions within technology education are two course sequences: Principles of Technology I and II, and Scientific and Technical Visualization I and II. The former is an applied physics program, and the latter focuses on skill development in graphic communications applied to technical and scientific concepts and principles.

There is statewide assessment (multiple-choice tests) for all technology education courses, with the exception of Exploring Technology Systems, Technology Advanced Studies, and the Project Lead the Way courses. The middle grades course, Exploring Technology Systems, is not part of statewide assessment, primarily because of its many permutations as used in the field. Local school systems are given the authority to use all or only parts of any middle grades course and also to determine the grade level (seventh or eighth) where it will be used. For instance, while the Exploring Technology Systems course is designed for 18 weeks of instruction (for seventh and/or eighth grades), many school systems choose to break the course up into smaller pieces. It may be offered in 6-, 9-, 12-, or 18-week increments during the seventh and/or eighth grades, depending on the instructional philosophy of the school system. Therefore, any assessment information gathered through a single statewide instrument would likely have low validity, and an attempt to design assessments to accommodate each of the possible variations of the course is beyond the finite resources of an educational system.

Technology Advanced Studies is designed to be an individualized project developed by the student and his or her teacher, and statewide assessment is not deemed feasible. Assessment is a local responsibility, and student outcomes are not reported to the state at this time.

Finally, Project Lead the Way courses are relatively new additions to the technology education program, are assessed locally, and are not currently part of statewide assessment. These courses will be reflected in the next *Standard Course of Study* (2009–2010), and in all likelihood will become part of the statewide assessment process at that time.

A BRIEF HISTORY OF THE TECHNOLOGY EDUCATION SCOPE AND SEQUENCE IN NORTH CAROLINA

In 1987, there were 23 courses within the technology education scope and sequence (two in middle grades and 21 in high school). Many of the high school courses had very low enrollment; some had none. Almost all course guides were little more than outlines suggesting a curriculum. Nearly all courses were adapted from work done outside North Carolina. In 1987, there was no statewide assessment. By 1991, the North Carolina Department of Public Instruction (NCDPI) decided to develop its own technology education curriculum and to dramatically reduce its course offerings. It was at about this time that North Carolina began its project to develop statewide assessments for most CTE courses, including those of technology education. Beginning in 1991, the NCDPI developed and released a new technology education scope and sequence that consisted of 13 courses. These courses evolved into the 11 that have been in place since 2002.

The Curriculum Development Process

For nearly every technology education course, the curriculum materials contain three primary components: (1) blueprints, (2) curriculum guides, and (3) banks of assessment items.

Blueprints may be thought of as course outlines made up of instructional objectives. Appearing on blueprints is information stating whether the objective is a cognitive concept (what a student should know), or a skill or performance statement (what a student should be able to do). Generally, technology education courses comprise approximately 50% cognitive concepts and 50% performance activities. Course blueprints also indicate the "weight" or relative importance of the concept or activity to the course as a whole.

The second component, the curriculum guide, gives explicit information about what a student is expected to know as well as detailed information about activities that support the objective, with accompanying assessment rubrics. The guide also contains a recommended equipment list, and references to literature used to develop the curriculum and facility plans.

The last component is the test-item banks, which are made up of multiple-choice questions and performance assessment items. There are two versions: one bank for classroom review and one secure. Teachers are given blueprints, curriculum guides, support materials (such as online presentations), and the corresponding classroom test-item bank on a CD.

Teachers are encouraged to use the classroom banks for interim reviews and assessments. Classroom banks generally contain 300 or more test items. Secure banks (one for each course) are used by NCDPI to assess virtually every high school student enrolled in every high school CTE course. These secure banks are held under tight security, and the teachers are not permitted access to them. They are similar, but not identical, to the classroom bank test items. (Readers may view the North Carolina technology education curriculum materials in their entirety at http://www.dpi.state.nc.us/workforce_development/technology/index.html.)

Assessment is an integral part of the curriculum development process. What is assessed is expected to be taught, and, conversely, what is taught is expected to be assessed. With this in mind, curriculum materials, which include blueprints, guides, test-item banks, and supporting materials, are developed by curriculum teams. Generally, a technology education curriculum team is led by a teacher educator and includes four to six practicing teachers who have a deep understanding and appreciation of the *Standards for Technological Literacy,* as well as expertise in the curriculum being designed.

During the course of development, drafts of the curriculum materials are shared with other teachers at the state's winter and summer conferences, permitting feedback and changes. Test-item banks of approximately 600 items (mostly multiple-choice) are developed by the curriculum team for each course as part of the curriculum development process. These items are then reviewed for content validity by teams of teachers, who each see no more than a third of the test items in any one bank. Two-hundred representative multiple-choice test items are then removed from the bank of 600 to become secure test items. During the first year the curriculum is released into the field, these 200 items are field-tested for reliability. Items that show poor reliability are removed from the secure bank. The secure test items with high reliability are used for statewide assessment until the next course iteration. Courses are revised approximately every five years. Performance assessment items are included in the curriculum guides, and it is expected that they will be used to develop and assess students' skills, activities, and products within the respective courses. Performance items have not been used in the statewide assessment, although some educators have argued for their inclusion.

Statewide assessment focused solely on students' cognitive understanding as measured by multiple-choice test items has raised questions that should be addressed:

1. Is the information gathered by the cognitive assessments a complete and accurate reflection of a student's ability to do what is desired of him or her?
2. Can he or she actually conduct a technology assessment, create a CAD drawing or video, or make an artifact?

3. Is this type of assessment an accurate measurement of the efficacy of a program?
4. Does focusing exclusively on students' cognitive outcomes change the environment and culture of the classroom in a negative or positive way?
5. Does it affect the program in a deleterious or a healthy way?

In an attempt to answer some of the above questions, a study was conducted during the 2004–2005 school year to assess the feasibility of including performance in the statewide assessment process. This project led to the development of performance assessment items aligned to specific objectives in selected CTE courses. Although performance assessments have long been a part of curriculum guides, many teachers seemed to consider them "optional" and did not see their connection to the multiple-choice tests upon which the accountability system is based. In the feasibility project, scoring rubrics were developed that clearly defined student outcomes and desired levels of performance. Teachers used these assessments in their classrooms and reported the results to the state on a spreadsheet. The teachers involved in the project were positive about the performance items, but less enthusiastic about how such an effort could be extended statewide. Their concerns included the time required for preparation, actual assessment, and scoring; the direct costs for assessment; issues of inter-rater reliability; and training required for the evaluators (Hoepfl, 2005).

The information gathered showed that most teachers wanted student performance assessment included in the statewide reporting of student outcome data. Participants suggested that over the years a significant number of teachers had changed their way of teaching, requiring students to spend more time focused on multiple-choice test-item, bank questions, and answers. This shift in focus was to the detriment of laboratory activities and complex cognitive tasks such as discussion, reading, and writing. Another concern developed when tentative evidence showed that in at least one non-technology education course (drafting), there was a lack of positive correlation between the cognitive and performance assessments.

In one drafting program, a third of the students who scored poorly on the cognitive statewide assessment did well on drafting performance assessments (drawings), and, perhaps more interestingly, a third of the students who did well on the cognitive assessments did poorly on the performance assessments.

There appears to be mounting evidence that the current statewide assessment process does influence the classroom environment and culture, and not always positively. The overall influence of the current assessment process upon program efficacy is more difficult to measure, but there is growing evidence that it has resulted in at least some harm. Nevertheless, neither teachers nor administrators argue to dismantle statewide assessment. Rather, there is the desire for improvement and, if feasible, to incorporate performance in the assessment model, as well as a more complete method of program assessment (Hoepfl, 2005).

ANALYZING AND USING DATA FROM THE STATEWIDE ACCOUNTABILITY SYSTEM

Academic Attainment

Figure 1 and Table 2 illustrate the performance of concentrators in technology education on the academic attainment measure from 2000–2001 to 2003–2004. In each of the four years for which data are available, a higher percentage of technology education concentrators exceeded the national average on the elementary algebra subtest than on any of the other three tests. This same pattern is also found with CTE concentrators as a whole. For example, in 2003–2004, among technology education concentrators, 48.1% of concentrators exceeded the national average in reading; in writing, 50.5%; in numerical skills, 57.0%; and in elementary algebra, 67.0%. Technology education concentrators showed small gains in their scores on three of the four subtests.

Aligning Curriculum and Assessment in the Development of a Statewide Accountability System

Figure 1. Academic attainment data for technology education students.

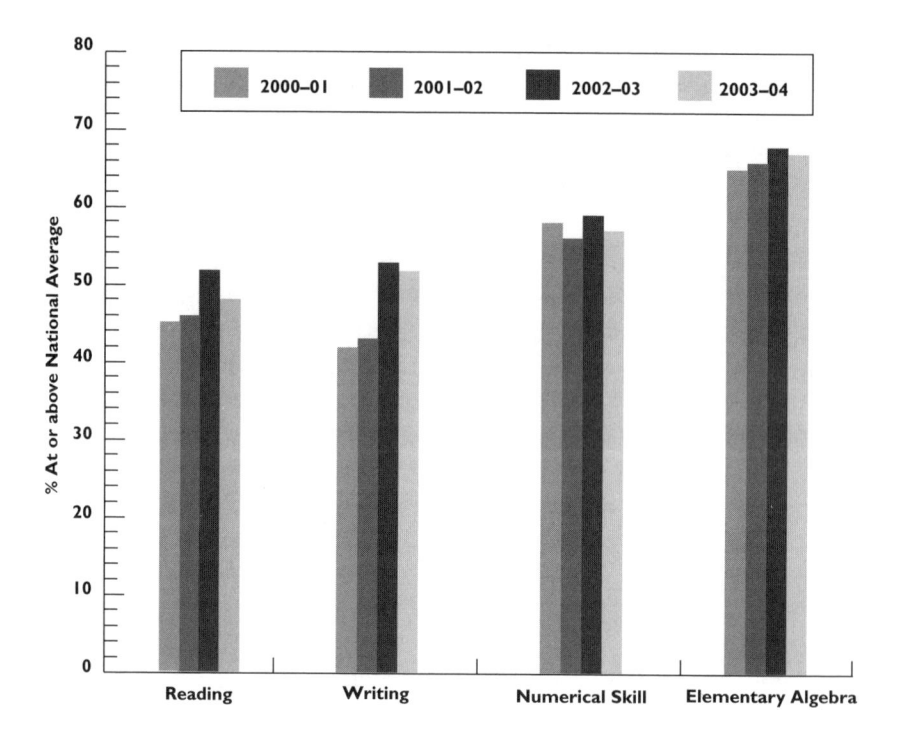

Table 2. Academic attainment in technology education.

Subtest	Academic Year			
	2000–01	2001–02	2002–03	2003–04
Reading	45.4	45.8	51.8	48.1
Writing	42.6	43.5	52.9	50.5
Numerical Skills	58.3	55.5	59.1	57.0
Elementary Algebra	65.7	65.9	67.8	67.0
Combined Scores	56.0	52.7	57.9	55.6

Technical Attainment

Figure 2 illustrates technology education students' performance on the technical attainment measure since 2000–2001. Details about overall and special populations technical attainment appear in tables 3 and 4. Since 2000–2001, the percentage of technology education enrollees who reached at least Level III proficiency increased from 41.0% to 48.0%, a 7.0% gain. Special populations students also showed a significant increase: from 27.2% at or above Level III in 2000–2001 to 34.7% in 2003–2004, a 7.5% increase. The technology education course with the highest performance on the technical attainment measure was Transportation Systems, where during 2003–2004, 58.1% of the enrollees scored at or above Level III.

Figure 2. Technical attainment scores for students in technology education courses.

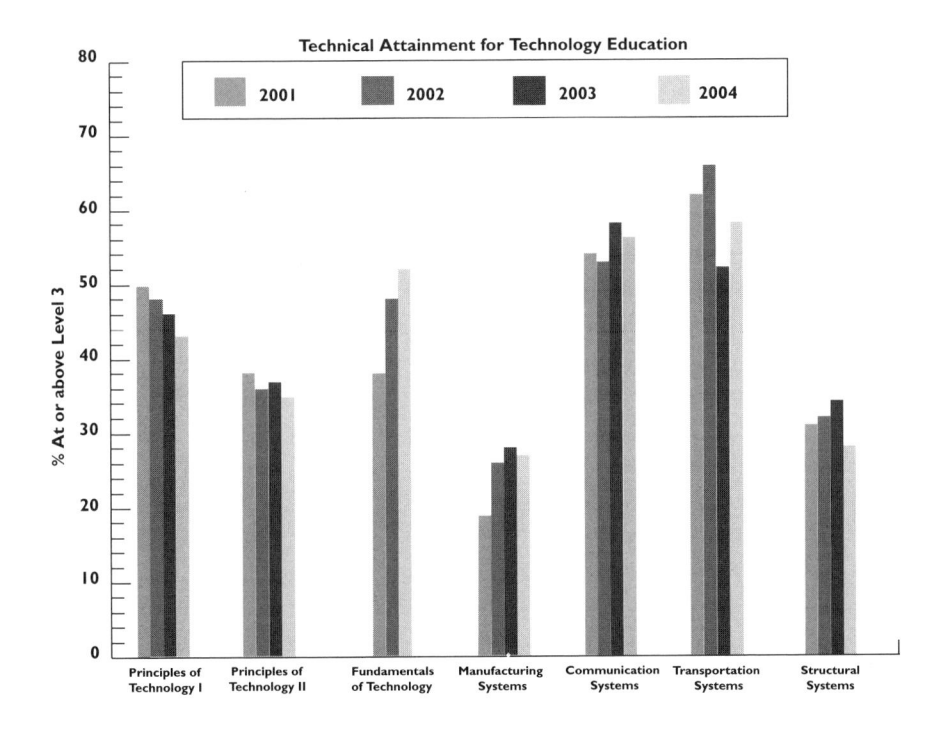

Aligning Curriculum and Assessment in the Development of a Statewide Accountability System

Table 3. Overall technical attainment by course in technology education: Percentage of students at or above Level 3.

Course		2000–01	2001–02	2002–03	2003–04
			Overall		
8011	Principles of Technology I	50.4	48.8	46.2	41.0
8012	Principles of Technology II	37.0	35.3	35.6	34.6
8010	Fundamentals of Technology	37.6	48.6	*	51.5
8015	Manufacturing Systems	19.4	25.2	28.3	26.6
8025	Communications Systems	54.7	54.6	58.7	56.3
8026	Transportation Systems	61.5	66.7	51.7	58.1
8041	Structural Systems	31.7	32.7	33.1	28.9
	Technology Education Total	41.0	47.5	45.9	48.0
	CTE Total	54.8	59.6	61.0	65.0

Table 4. Special populations technical attainment in technology education: Percentage of students at or above Level 3.

Course		2000–01	2001–02	2002–03	2003–04
			Special Populations		
8011	Principles of Technology I	36.1	34.2	27.7	26.2
8012	Principles of Technology II	25.6	22.2	18.3	20.0
8010	Fundamentals of Technology	22.4	32.3	*	37.0
8015	Manufacturing Systems	11.5	13.7	14.4	17.6
8025	Communications Systems	46.3	42.8	44.4	45.2
8026	Transportation Systems	41.6	51.3	45.6	50.0
8041	Structural Systems	21.5	21.4	19.5	20.2
	Technology Education Total	27.2	32.4	31.2	34.7
	CTE Total	41.9	47.4	45.9	54.5

Table 5 shows the average score by course in technology education. There were no significant changes in state averages for technology education courses between 2002 (the first year this information was available) and 2004.

Note that gaps in the charts are likely due to introduction of new curriculum. When a new curriculum is introduced, there is a one-year gap in reporting testing data, during which updated test items are validated and reliability is determined.

Table 5. Average score by course in technology education.

Course		Academic Year		
		2001–02	2002–03	2003–04
8011	Principles of Technology I	63.1	62.6	60.1
8012	Principles of Technology II	57.1	57.2	57.2
8010	Fundamentals of Technology	62.2	*	62.6
8015	Manufacturing Systems	54.6	56.3	55.0
8025	Communications Systems	64.8	66.2	65.4
8026	Transportation Systems	69.1	64.8	66.4
8041	Structural Systems	56.8	56.7	56.8

FOCUSING FUTURE EFFORTS

Developing curriculum based on desired student outcomes and using that as a model to assess technical attainment will certainly continue in North Carolina. However, changes in the political climate may have an impact on assessments and how they are used.

Although nationally CTE is specifically exempt from provisions of the No Child Left Behind Act, such efforts will undoubtedly play a part in determining the future role of accountability in CTE. Predictions are that the Perkins legislation, awaiting reauthorization as of this writing, will strengthen accountability requirements for CTE. However, there is a risk that the overwhelming testing and data analysis requirements of NCLB may result in a backlash against testing, and CTE could certainly be caught in the fallout. There are already many teachers in North Carolina who, although very supportive of the VoCATS curriculum materials, hesitate to support the testing component. In spite of numerous promises to the contrary, teachers perceive student test results as being used to assess teacher performance by local administrators.

In a promising direction for VoCATS, North Carolina is moving toward a more complete assessment model, as previously discussed, which would include performance assessment, where appropriate, in addition to multiple-choice tests. This project will continue, and the state education department staff will continue to look for ways to make the state accountability effort "more authentic."

Another challenge facing curriculum developers is the difficulty in maintaining a state-of-the-art curriculum, particularly in areas that include technology. For some CTE courses, North Carolina has already had to drop state-developed curriculum in favor of commercially available materials. These materials frequently include assessment measures, but because of their proprietary nature, results may not be made available for use in the accountability system. In addition, there is no way to equate materials and tests from different vendors. Local administrators, therefore, frequently request permission to use results from national certification examinations in lieu of post-assessments. However, due to the prohibitive cost, local districts cannot require students to take the certification exams, and reporting results only for those students who thought it was worth the expense to take the exam are not representative. Incorporating national standards into the accountability system is also a challenge. Which standards should be selected for each course? If there are competing groups of standards, how do developers determine which groups' standards to include?

The cost of developing curriculum and assessments at the state level, of administering assessments and analyzing data at the local level, and of collecting and analyzing data at the state level are also prohibitive. Since the curriculum development and assessment processes are fully integrated, it is not possible to determine exactly how much is spent annually on assessment, but it is certainly substantial. Officials will need to demonstrate the value of accountability to ensure that funds continue to be available for this purpose.

CONCLUSION

In spite of the expense (human and capital), the complexity, and the problems encountered in the current statewide curriculum development and assessment process, its value far exceeds its costs. In 1990, it was impossible for teachers or local and state administrators to make claims of certainty regarding the quality of their programs. There were no measures of student outcomes other than those a teacher chose to use in his or her classroom or upon which judgments could be based. There was not a well-developed common language that teachers and administrators shared, there was no public debate regarding what constituted a well-defined, quality curriculum, and the *Standards for Technological Literacy* had yet to be written.

Well-defined student learning outcomes are predicated on well-defined curriculum, tools, and facilities. Before VoCATS, teachers were left to their own devices and the good will of administrators for providing quality facilities. Statewide standards now provide these guides. This same principle also has had an ameliorating influence for addressing the needs of low-income communities. It is difficult for state officials to pronounce, on the one hand, "No Child Left Behind" and, on the other hand, not provide the needed resources (as identified in program requirements) that all children need in order to succeed.

The United States has historically left curriculum development, implementation, and assessment in the hands of the classroom teacher or the local school system. If comparisons between other industrialized nations are to be believed, the United States is failing by most measures in reading, writing, mathematics, and the sciences. What claims can be made with regards to technology? North Carolina has demonstrated that focused, coherent standards and assessment provide a basis upon which to answer that question.

DISCUSSION QUESTIONS

1. Should there be formal standards for technology education performance, or should individual teachers develop their own standards? Explain your opinion. If you believe there should be standards, who should set and monitor them? What are some inherent problems with establishing external standards on a state or national basis?
2. Can an accountability system without a performance assessment component, and relying on paper-and-pencil assessments only, provide an accurate picture of student achievement? Why or why not? How could performance assessment be included in this system?
3. How accurately do student scores on a statewide assessment reflect the quality of an instructional program? What factors other than the teacher and the curriculum influence student scores?
4. How can a statewide system of curriculum and assessment be responsive to changes in a technology-based field to maintain the state of the art?

REFERENCES

Bloom, B.S., D. R. Krathwohl, D.R., and B. B. Masia (1984). *Taxonomy of educational objectives: The classification of educational goals.* New York: Longman.

Hoepfl, M. (2005). *NCDPI career and technical education performance assessment feasibility project: External evaluation final report.* Raleigh, NC: North Carolina Department of Public Instruction.

North Carolina Department of Public Instruction (2002). *Standard course of study guide.* Raleigh, NC: North Carolina Department of Public Instruction.

Developing a Large-Scale Assessment of Technological Literacy

Chapter 9

Stephen Petrina
University of British Columbia

Ruth Xiaoqing Guo
University of Ottawa

INTRODUCTION

In *Monty Python and the Holy Grail,* King Arthur sets off in quest of the elusive and coveted treasure. However, in one of the first scenes, after confronting a French-occupied castle and inviting the occupants to join him on his quest for the Holy Grail, he gets flustered after being told that the castle already has one, and the occupants need not foolishly spend any time on the quest. Thoroughly defeated by the taunting of the French, Arthur decides that the best way to successfully conclude the quest is to do it individually, rather than as part of a group. Albeit without the bumbling and laughs that Monty Python inspired, this is the decision that technology education researchers now face: continue individual searches for the Grail of a Technological Literacy Scale or collectively work through agencies such as the National Academy of Engineering (NAE) and Educational Testing Service (ETS). In this chapter, the Grail refers to a norm-referenced, standardized, universally applicable scale of technological literacy (Leach, 1997). Our task would be simple if only the politics of large-scale measurement were given comic relief by Monty Python and limited to cooperative quests for the Grail. More realistically, the politics of large-scale assessment raise questions of access to high-stakes knowledge and about the standardization of curriculum, or the alignment of the "attained" curriculum with the "intended" curriculum. More specifically, large-scale assessments raise concerns about generating norm-based references for pitting students against students, schools against schools, and countries against countries. Conceivably, valid, norm-based scales of technological literacy could generate this competitive scenario. On the other hand, one of the hallmarks of high-status subjects (e.g., English, math, and science) is the use of standardized scales for large-scale comparisons. Indeed, trading off locally controlled, culturally specific engineering, design, and technology curriculum and assessments for internationally recognized scales of technological literacy could provide the high status that so many educators and researchers desire. After all, one path to high status and the Grail was paved by the International Technology Education Association (ITEA), with its release of *Standards for Technological Literacy* (2000).

On one level, the standards offer a clear definition of the construct under question: Technological literacy is merely "the ability to use, manage, assess, and understand technology" (ITEA, 2000, p. 9). The standards provide clear content domains for constructing a bank of items to select for an instrument, and scales for measurements and comparisons of technological literacy could easily be created. This step was more or less taken by the ITEA in *Measuring Progress: A Guide for Assessing Students for Technological Literacy* (2004). With the addition of a performance assessment component, the Grail of a Technological Literacy Scale may be in reach. On another level, however, technological literacy is a deeply contested construct—a fact that was underappreciated by the architects of the standards (Cope and Kalantzis, 2000; Dakers, 2006; Keirl, 2006; Petrina, 2000a, 2000b, 2003). Coincidental with the ITEA's release of the *Standards for Technological Literacy* was the International Society for Technology in Education's (ISTE, 2000) *National Education Technology Standards for Students (NETS)*. Similarly, in 2002 the ETS signaled its interest in measuring technological literacy with the release of *Digital Transformations: A Framework for ICT Literacy,* and in 2004 ETS unveiled the first standardized instrument for large-scale assessments of information and communication technology (ICT) literacy. However redundant, the prevalence of terms such as "engineering literacy," "digital literacy," and "media literacy" further muddy the waters. Science, technology, and society (STS) researchers and environmentalists also offer constructive positions on technological literacy that cannot be overlooked. The construct of technological literacy differs across agencies, and there are no shortages of disagreements over whether large-scale assessments ought to address functional literacy or critical literacy, or strike a balance between the two. Despite this, there is a consensus that technological literacy, however defined and operationalized, is an extremely important construct and increasingly recognizable among competing and multiple literacies (Cope and Kalantzis, 2000; Kahn and Kellner, 2005).

In this chapter, by exploring the relative complexity of the landscape for conducting large-scale assessments of technological literacy, we address the challenges in creating and sustaining these assessments. Some of the better-known large-scale measures, such as Pupil's Attitudes Toward Technology (PATT), are sustainable for explicit reasons, including an open-source philosophy. Open source means that the instrument (or software, information, etc.) is freely circulated and is open to customization, translation, redistribution, or anything but commercial, profit-driven appropriation. Other efforts, such as the British Assessment of Performance Unit, amassed a volume of expertise but were resource-intensive and unsustainable. We describe a few smaller-scale but noteworthy efforts as well and caution against rejecting open-source instruments and graduate research studies in favor of proprietary instruments and sponsorship by the ETS or NAE. We argue that conventional deficit models (i.e., those that hold that students are deficient in knowledge, skill, or other attributes) tend to govern research designs and the development of assessments, and ultimately simplify interpretations of technological literacy data. Without taking critiques of deficit models into account (e.g., Bak, 2001; Irwin and Wynne, 1996), theorizing *literacies* as having social dimensions (Cope and Kalantzis, 2000), or understanding technology as a social process, we risk basing normative judgments on deficit models and thus limiting technological literacy to problems with, and properties of, individuals. This chapter is partially a review of large-scale assessments of technological literacy and partially an analysis of issues and problems that threaten the legitimacy of these assessments.

LARGE-SCALE ASSESSMENTS OF TECHNOLOGICAL LITERACY

One of the most ambitious large-scale assessments to date is the Third International Mathematics and Science Study (TIMSS), which took six years from development in the early 1990s to implementation in 1995 and 1999, and involved 1.3 million fourth- and eighth-grade students in 49 countries (Howie, 1999; Martin and Kelly, 1996). By any stretch of the imagination, TIMSS was large-scale, and the implications for curriculum, policy, and teaching were far-ranging. Hundreds of paper and pencil items were constructed for the tests, and performance tasks were created for hands-on problem solving. Performance assessment was limited to nineteen countries at the eighth-grade level and nine countries at the fourth-grade level. This was no mean feat, since manipulatives had to be standardized and distributed en masse to researchers overseeing the assessments in individual schools and classrooms across the world (Robitaille et al., 1997). One task challenged the students to judge and compare charges of batteries, while another involved calculations for sizing furniture that could be potentially moved around a corner in a hallway of specific dimensions (Garden, 1997). Coordinators were often eager to note that the assessments addressed technological literacy (Orpwood and Garden, 1998), and TIMSS 2007 promises to involve an even larger measure of technological knowledge and problem solving. Indeed, economies of scale suggest that assessment specialists in technology education would do well to link technological literacy to TIMSS 2007.

Derived from international measures in the 1960s, TIMSS is an artifact of the problem of status in the curriculum. Since that time, governments and educational agencies have been anxious to respond to large-scale measures of math and science but less interested in results of assessments in low-status subjects. The National Assessment of Education Progress (NAEP), which began in 1969 and is specific to the United States, conducts large-scale assessments in 11 subjects (including civics, economics, foreign language, geography, mathematics, reading, science, U.S. history, world history, and writing). While the NAEP involves a mix of high- and low-status subjects, technology is not among them (e.g., Kendall and Marzano, 1997). One must turn internally to the professions of technology education and STS for large-scale assessments of technological literacy.

In the mid 1980s, Jan Raat and Marc de Vries of Eindhoven University created the PATT instrument for assessing attitudes, values, and general understanding of concepts. By 1987, the instrument was used for surveys in 20 countries (Raat et al., 1987). PATT was transformed for the U.S. context in 1988 when de Vries, along with Allen Bame and William Dugger of Virginia Tech, designed a large-scale assessment of middle school students (Bame et al., 1993). A total of 10,269 students from 128 schools in seven states participated in this first large-scale PATT U.S.

assessment. PATT continues to be popular and, across the world, remains one of our most reliable tools for comparative measures of students' attitudes and values toward technology. The PATT instrument typically consists of 100 Likert-type items. The first 11 items are used to collect demographic information from the individual students. The remaining items deal with issues that force students to form an opinion (there is also a "neutral" option) or express their understanding of concepts. Sample items include the following:

Agree   Tend to Agree   Neutral   Tend to Disagree   Disagree

12. When something new is discovered, I want to know more about it immediately.
18. I would like to know more about computers.
24. A girl can become an auto mechanic.
40. I think visiting a factory is boring.
43. To study technology you have to be talented.
69. With a technological job your future is promised.
76. In my opinion, technology is not very old.
97. Technology has little to do with daily life.

This is similar in design to scales of environmental values, militarism-pacifism, and the politics of technology. However, these latter scales tend to be inherently more political.

The PATT questionnaire discriminates between students who have and have not taken technology courses. We have learned from PATT that taking a technology course makes a "significant difference" in both outlook toward, and knowledge of, technology (Bame et al., 1993, p. 46; see also Ankiewisc, van Rensburg, and Myburgh, 2001; Becker and Maunsaiyat, 2002; Boser, Palmer, and Daugherty, 1998; van Rensburg, Ankiewisc, and Myburgh, 1999; Volk and Ming, 1999; Volk, Yip, and Lo, 2003). Another important finding in the PATT assessments is that attitudes toward technology, as we might expect, rise and fall with economic outlooks and realities (Volk, Yip, and Lo, 2003). Nonetheless, PATT is a paper-and-pencil instrument for providing measures of attitudes and general understanding of concepts, considering that attitudes and understanding of concepts are only two dimensions of technological literacy.

British design and technology (D&T) researchers were among the first to conduct large-scale assessments of technological literacy, including a significant performance component (Kimbell et al., 1991). Directed by Richard Kimbell and the Assessment of Performance Unit (APU) at Goldsmiths College, the 1988–89 D&T assessment generated 20,000 artifacts—design brief explanations, drawings, portfolio entries, and so on—from about 10,000 students and 700 schools, and required 120 raters to deal with the evidence (Kimbell, 1997, pp. 28–43; see also Ch. 10 of this yearbook). Concentrating on the process of design, or design cognition, in the performance component of the assessment, the APU attempted to provide norms of ability or design cognition for age groups (Kimbell, Stables, and Green, 1996, pp. 48–86). The APU's time-consuming, nuanced assessments provide researchers with indicators of attainment and progression from one level of capability and literacy to another. The APU helped establish D&T as a compulsory subject in the National Curriculum of England and Wales beginning in 1990, but their assessment philosophy (i.e., holism) contradicted the government's Task Group on Assessment and Testing's philosophy (i.e., atomism).

In 1999, nearly 3,500 students in grades five, eight, and 11 in 182 Saskatchewan schools participated in an assessment of technological literacy, which generated about 7,000 artifacts to assess learning (Saskatchewan Education, 2001). Similar to the APU's work in England, this large-scale assessment combined the *Views on Science-Technology-Society* (VOSTS) questionnaire, developed at the University of Saskatchewan (Aikenhead and Ryan, 1992; Aikenhead, Ryan, and Fleming, 1989), with a performance component. The assessment addressed four dimensions of technological literacy:

1. understanding, describing, and adapting technology
2. accessing, processing, and communicating information
3. responsible citizens and technology
4. using technology, including computers

For the most part, the results were disconcerting. Saskatchewan students generally met standards for the first dimension but fell short of standards expected for the remaining three dimensions (Saskatchewan Education, 2001, pp. iii–v).

Obviously, research into how we become technologically literate requires an array of evidence, some conversational and some documentary- and performance-based, depending on the instruments and methods used. One conclusion to draw is that researchers are entering classrooms with complex theories about learning *and* normative expectations of student performance levels. Learning theorists and researchers are not necessarily working in tandem with large-scale assessment specialists, but there are levels of articulation between the two.

Outside of schools, research into the public understanding of science and technology has attracted the interests of policy makers, and researchers in science, technology education, and STS have been generally responsive to demands for information. In these types of large-scale assessments, economic agendas often run up against activist-researchers interested in contradicting overemphases on market growth and progress. Government or industry-sponsored large-scale assessments are often oriented toward defending or increasing capital investments in fields such as engineering and science (e.g., Miller, 1986, 1992), while activist-researchers orient their assessments toward defending the environment, labor rights, or sustainability (e.g., Kempton, Boster, and Hartley, 1996). This is partially what makes social science surveys both interesting and suspect.

The U.S. National Science Foundation (NSF) sponsored five large-scale assessments of the public understanding of science and technology in 1979, 1981, 1985, 1988, and 1990 (Miller, 1986, 1992). For the 1990 assessment, 2,033 adults (from 33 cities and 117 rural areas) were contacted by telephone by the Public Opinion Laboratory at Northern Illinois University (Miller, 1992). Each adult was first asked if they were knowledgeable about specific issues regarding science and technology. If they responded no, then these items were omitted from the survey. With a response of yes, the researchers progressed to corresponding items (about 150 in total). The repeated measures design allowed the researchers to interpret trends on particular items over the five surveys. For example, responses were similar (i.e., 43% disagreed) across all of the surveys to the item "We depend too much on science and not enough on faith." Similarly, over time respondents consistently agreed (85%) with the item "Science and technology are making our lives healthier, easier and more comfortable" (Miller, 1992, p. 43). This somewhat consistently enthusiastic outlook toward science and technology contradicts large-scale survey results and much more wary public attitudes from 1972 (La Porte and Metlay, 1975). Not surprisingly, the participants in the 1990 survey were invariably interested in environmental issues (e.g., acid rain, ozone hole, nuclear power, pollution). As one indicator, 76% of the participants felt that the level of government spending in reducing pollution was "too little" (p. 71). Overall, only 7% of the sample was considered to be scientifically and technologically literate (pp. 14–15).

Ecology and technology are inseparable—technological literacy is necessarily environmental literacy (Elshof, 2003; McLaughlin, 1994). Hence, in large-scale assessments of environmental literacy we can derive an understanding of technological literacy. Kempton, Boster, and Hartley's (1996) NSF-funded survey is a textbook example of these interconnections and of the politics of social science surveys. For instance, their *Environmental Values* instrument includes the following items (150 total items):

Strongly Agree	**Agree**	**Slightly Agree**	**Slightly Disagree**	**Disagree**	**Strongly Disagree**

15. We're advancing so fast and are so out of control that we should just shut down and go back to the way it was in colonial times.
23. People should pay the environmental costs of the things they buy. Products should be taxed depending on their effect on the environment.
37. We don't have to reduce our standard of living to solve global climate change or other environmental problems.
117. As new technologies become available that are less environmentally damaging, companies will naturally want to adopt and use them.

Kempton, Boster, and Hartley's survey is large-scale, not in its number of participants (n=142, 33% female), but in the scope of the research design and instrument. They sampled from target populations of members in Earth First! and the Sierra Club, along with dry cleaners, sawmill workers, and laypeople. Values, as they demonstrated, are intricately tied to identities. On items dealing with intrinsic rights of nature, sawmill workers were the least likely to subscribe to biocentric values. Perhaps not surprising, 97% of the Earth Firsters and 63% of the sawmill workers agreed with the following: "Justice is not just for human beings. We need to be as fair to plants and animals as we are towards people" (p. 113). This type of differentiation within items is crucial to interpretation and to understanding the politics of large-scale assessments. Rather than avoid the politics inherent in technological literacy, Kempton, Boster, and Hartley's strategy was to embrace and interpret these politics.

One of the more recent public understanding surveys was the ITEA/Gallup Poll, administered in March and June of 2001. The ITEA/Gallup Poll surveyed a cross-section of American adults (n=1,000, 48% female) on 17 items that represent technological literacy and the support of technology as a subject (Rose and Dugger, 2002; Pearson and Young, 2002; Petrina, 2003). Three quarters of the respondents considered themselves able to adequately use and understand technology, as basically a residue of incidental experiences with technology. However, only 53% could explain how energy is transformed into electrical power. This seemingly trivial knowledge about everyday technologies might mean the difference between supporting alternative energy policies and settling for status quo fossil fuel-fired electrical generation plants. The adults sampled were nearly unanimous in agreeing that technological literacy was somewhat or very important, but 67% (78% of the 18–29 age group) associated technology with computers (Rose and Dugger, 2002, p. 2).

Large-scale assessments of computer literacy, which is now called digital or ICT literacy, has a history similar to that outlined above. Computer literacy came of age in the late 1970s and early 1980s at the dawn of the microcomputer revolution. Like technology educators, computer educators struggled with a balance of applications and implications in their large-scale assessments. In computer literacy, British Columbia technology educator Annette Wright lamented in 1980, "the stress would undoubtedly appear to be on the technical and mechanistic aspects of computers, to the detriment of their sociological aspects—privacy, security, convenience, learning modes and problem solving" (p. 8). Instruments of computer literacy typically emphasized an individual's perception of competencies or skills and self-efficacy (e.g., Albion, 2001; Feng, 1996; Fisher, 1997; Jones and Pearson, 1996; Kellenberger, 1996; Knezek and Christensen, 2000; Oderkirk, 1996; Woodrow, 1991). Attitudinal measures tended to focus on computer phobia or anxiety and, theorized through deficit models, were intended to help children or adults overcome their so-called fear of computers (Cambre and

Cook, 1985; Heinssen, Glass, and Knight, 1987; Weil and Rosen, 1997). *The Children's Attitudes Toward Technology Scale* (CATS) and the *Young Children's Computer Inventory* are good examples of this (Frantom et al., 2002; Knezek, Miyashita, and Sakamoto, 1994).

One of the largest assessments of computer literacy was sponsored by the International Association for the Evaluation of Educational Achievement (IEA) from 1987 to 1993 (Pelgrum et al., 1993). The IEA administered a 30-item self-rating Functional Information Technology Test (FITT) to 7,092 fifth-grade students in the United States and the Netherlands and to 23,102 eighth-grade students in seven countries, including the United States and the Netherlands. FITT scores were highest for Austria and Germany, where an emphasis is placed on understanding the way software and hardware work. Like many of these studies, data for items dealing with access to computers and other devices in and outside of school are obsolete before executive reports are published. This was a problem that plagued standardized tests of computer literacy, such as the battery of Computer Literacy and Computer Science Tests, New Technology Tests, and the Standardized Test of Computer Literacy and Computer Anxiety Index (Conoley and Kramer, 1989). Neither *The Sixteenth Mental Measurements Yearbook* (Spies and Plake, 2005) nor *Tests in Print VII* (Murphy et al., 2006), the latest compilations of standardized tests, contains contemporary computer or technological literacy instruments that meet Buros Institute of Mental Measurements' primary criterion for inclusion (i.e., minimal instrument design information is available). This could be a sign that, as in *Monty Python and the Holy Grail*, everybody already has one or, alternatively, that nobody wants one or that the Grail will be found in the future.

Our large-scale survey of ICT literacy at the University of British Columbia (UBC) utilized a locally made, self-efficacy instrument to assess competencies and attitudes (Guo, 2006). We administered pre-program and post-program instruments to teacher education students in 2001 and 2002 and again in 2003 and 2004 (n=2874, 70% female). We revised the post-program instrument to eliminate items that did not discriminate and replaced them with disposition items informed by critical theories of ICT literacy. These Likert items included statements such as: "The World Wide Web advances gender and racial equity," "Significant online game playing (i.e., two hours or more per day) promotes hyperactive, aggressive behavior," and

"Information technologies are just tools." Responses to the items were volatile, with sidebar comments such as "absurd," "bizarre," "bogus," "gendered," "sexist," "inappropriate," "impossible to answer," "offensive," "strange," and "opinion, not fact-based." A number of students requested an "I don't know" or "neutral" option for these items. One student wrote: "without reading studies on these statements it is difficult to form an opinion." These types of data are priceless, and had we been unable to customize the instrument, we would have overlooked many of the students' reticence toward critical literacy. We traded generalizability and external validity for flexibility. Another limitation of this assessment of ICT literacy was that we did not employ performance tasks.

Standardized, proprietary instruments offer the advantages of norm-referenced assessment, providing levels of reliability and validity that more local instruments fail to establish (Ghiselli et al., 1981). It was not surprising when, in 2004, the ETS signaled its interest in ICT literacy. In November 2004 the ETS unveiled its ICT Literacy Assessment, a 75-minute online instrument consisting of 13–15 scenario-based tasks. The tasks challenge participants to utilize their knowledge and skills of ICT to analyze spreadsheets, browse for specific information, configure databases, create graphs, and map concepts, among a variety of other outcomes. These are the types of competencies typically associated with "information fluency," a construct popularized by cybrarians, librarians, and information scientists (e.g., Committee on Information Technology Literacy, 1999). For the assessment, the ETS (2002) defines the construct as follows: "ICT literacy is using digital technology, communications tools, and/or networks to access, manage, integrate, evaluate, and create information in order to function in a knowledge society" (p. 2). Similar to the architects of *NETS,* the ETS defines ICT literacy as "functional or instrumental" literacy (see also Eshet-Alkalai, 2004). In January 2006, an advanced ICT Literacy Assessment was launched. Large-scale piloting began in the spring of 2005 when 3,000 students in the California State University system participated. The pilots established baselines and norms for standardization.

One could argue that the ETS's Praxis II: Technology Education (0050) test for teacher certification is also an assessment of technological literacy. Roughly 70% of the assessment deals with ICT, construction, manufacturing, and energy/power/transportation. The Praxis tests are a good example of the politics of assessment. Their purpose is for gate-keeping, but there is no evidence of their connection to the improvement of the teaching force in the United States (Angrist and Guryan, 2004).

CAUTION: DEFICIT MODELS AND LARGE-SCALE ASSESSMENTS

Important interpretations can be derived from large-scale assessments, but science and technological literacy theorists caution against basing normative judgments on deficit models and limiting technological literacy to problems with, and properties of, individuals (Fourez, 1997; Roth and Lee, 2002). Deficit models suggest that individual students and the public are basically inadequate in their knowledge of, and lack appropriate values toward, in our case, technology. It is important to recognize that literacy is collective as much as it is individual. The "real" deficits for illiteracies and inadequacies are likely to be found in policies.

Although education is normative by definition, deficit models have limitations. For example, anti-racist, feminist, Marxist, postcolonial, and queer theorists detail the problems of educating students in heteronormal, privileged, white institutions. What policy makers, researchers, or teachers perceive as a deficit in students may actually be a deficit in policy, equity, or acceptance of difference. Hence, identifying and filling up cognitive, affective, or disciplinary deficits can be a damaging practice. Researchers caution that problems of difference, inclusion, or equity are not merely functions of interpretation but are built into curriculum and assessment instruments (Petrina, 2001). In science and technology studies, researchers found that measures of scientific and technological literacy were used to justify increases in resources for science and technology (e.g., Bak, 2001; Irwin and Wynne, 1996). Deficits in science and technological literacy also justified increases in resources for putting a positive spin on controversial issues, such as biotechnology and nuclear power, to counter "ignorance" (Bak, 2001; Irwin and Wynne, 1996). Also, with research suggesting that little has changed in uses of ICT in public school classrooms and in academic achievement over the past 20 years, despite billions of dollars of investments and lucrative partnerships for corporate vendors, administrators, and investors are searching for easy scapegoats (Cuban and Kirkpatrick, 1998; Waxman et al., 2003). Blame is placed on classrooms instead of boardrooms, and on literacies instead of policies. Deficits of technological literacy become justifications for more of the same policy: more capital resources and more revenues for corporate vendors. Research and interpretation of technological literacy data have to be sophisticated enough to accommodate activism and resistance to market-oriented arguments for more capital investments at the expense of equity, environmental justice, and human rights.

CLOSING IN ON THE GRAIL?

One influential factor that helped sustain the PATT instrument for over twenty years is its open-source philosophy. Also, from its inception PATT connected assessment with dissemination and interpretation of research through its annual and bi-annual meetings. The relatively free diffusion of the instrument across the world provided a rich resource of data for comparative analyses. In the post hoc aggregate, PATT is a very large-scale assessment, but PATT began as a fairly small experiment in social science research. Disaggregated into hundreds of researchers and studies, PATT is only one of a variety of notable small-, medium-, and large-scale efforts to assess technological literacy.

One of the earliest questionnaires for technological literacy was Blomgren's (1962) Understanding American Industry Test. Administered to 151 freshman and seniors (all males) at Illinois State Normal University, the instrument embodies a sophisticated analysis of the politics of literacy and technology. The graduate research of Richter (1980), Hameed (1988), Hayden (1991), Kuforiji (1992), and Welty (1992), which involved the development or implementation of instruments for measuring technological literacy, are noteworthy as open-source and accessible. Siciliano and Maser (1997) also developed a helpful instrument, *Disciplinary Understanding and Attitudes Toward Technology,* for assessing technological literacy. Ideally, designers of large-scale assessments would come to terms with these efforts and collate items and performance problems in a bank from which one could generate instruments and scales. This would also mean coming to terms with computer and digital literacy assessments and their relationship to a broader measure of technological literacy.

The entry of the ETS into the quest for an assessment of technological literacy was predictable. Vendors of proprietary goods recognize incipient and changing markets. There is money to be made in the quest for measures of technological literacy. Private, for-profit educational companies have been flourishing in this market for technological training and welcome opportunities to assume the task of large-scale assessment and certification (Petrina, 2005). Vendors have the advantage of committing relatively large volumes of resources to the development of marketable instruments. However, limited access to standardized instruments is a problem. TIMSS and its counterpart, the Progress in International Reading Literacy Study (PIRLS), are sustainable primarily because of national and international research funding agencies' interest in high-status subjects. Advantageously, the large bulk of funds for TIMSS and PIRLS flow to university researchers and not to private corporations. Nevertheless, quantifiable research prevails over narratives of literacy, including technological literacy (Hoepfl, 1997; Selfe and Hawisher, 2004; Zuga, 1987).

For technological literacy theorists and assessment specialists, the entry of the NAE into the quest is an extremely promising sign. Although there was an initial under-theorizing of technological literacy in *Technically Speaking,* the NAE's platform for encouraging large-scale assessment, some progress has been made in overcoming these oversights (Garmire and Pearson, 2006; Pearson and Young, 2002; Petrina, 2003). For the purposes of this chapter, the key recommendation of the NAE was this: "The National Science Foundation (NSF) should support the development of one or more assessment tools for monitoring the state of technological literacy among students and the public in the United States" (Pearson and Young, 2002, p. 109). With the realignment of engineers with technology education comes the reintegration of engineering with technological literacy (Lewis, 2004, 2005; Robinson and Kenny, 2003; Rogers, 2005). It is nevertheless easy to exaggerate or erroneously extrapolate from short-term trends. Many technology educators will remember the enthusiasm over the alignment of design with technological literacy and the signaling of the Industrial Designers Society of America's (IDSA) intent to join with the ITEA on the quest for the Grail in the early 2000s. Yet for all the ceremony that accompanied the IDSA's joining of the quest, we are no closer to the Grail now than we were before. Or are we?

In *Monty Python and the Holy Grail,* the movie and the quest end precisely where they started: at the French castle where the Grail supposedly resides. Having gone it alone, those seeking the Grail are once again united, and once again the French refuse to give up the Grail and are instead thoroughly insulting to those at the gates. The individual efforts failed to satisfy the quest. Collectively storming the castle remains an option, but no one has the stomach for defending against the taunting. It may very well be that, as Arthurian and Grail romances scholar Joseph Campbell (1990, pp. 209–227) points out, the quest will not end with the Grail of a Technological Literacy Scale in hand. The challenge is to be on the quest, not to conclude it. In concluding the quest for a Scale of Technological Literacy, we risk premature closure on a measure to capture the dynamic nature of technology and technological literacy.

The time is right for someone to offer a Third Way for assessments of technological literacy. Where large-scale efforts offer the benefit of standardization (i.e., reliability and validity), inferential measurement of individuals and small-scale efforts offer a benefit of customizability for local nuance, performance assessment, and narratives. A Third Way might mediate between and balance the two with an emphasis on collective technological literacy. A Third Way might omit individual assessments in favor of social group assessments and accommodate for a quantification of team performance and the qualification of collective stories of growing up in a contradictory and increasingly technological world. A Third Way might shift from individual snapshots of capability to a greater understanding of the collective process of becoming technologically literate over the lifespan (Petrina, Feng, and Kim, in press).

DISCUSSION QUESTIONS:

1. What are the challenges in creating and sustaining a large-scale assessment of technological literacy?
2. What are the lessons to be learned from TIMSS and other existing large-scale assessments?
3. What must be done to elevate technology education to "high status" within educational circles so that large-scale assessment of technology education becomes desired?
4. Who should take on the task of creating a large-scale assessment of technology education?

5. Why is the quest for the Grail of a large-scale assessment of technological literacy as or more important than the assessment itself?

REFERENCES

Aikenhead, G., and A. Ryan (1992). The development of a new instrument: Views on science-technology-society (VOSTS). *Science Education* 76:477–492.

Aikenhead, G., A. Ryan, and R. Fleming (1989). *Views on science-technology-society*. Saskatoon, SK: University of Saskatchewan.

Albion, P. (2001). Some factors in the development of self-efficacy beliefs for computer use among teacher education students. *Journal of Technology and Teacher Education* 9(3): 321–347.

Angrist, J. D., and J. Guryan (2004). *Does teacher testing raise teacher quality? Evidence from state certification requirements.* http://www2.gsb.columbia.edu/divisions/finance/seminars/micro/spring_04/teachertest.pdf (accessed January 16, 2006).

Ankiewisc, P., S. van Rensburg, and C. Myburgh (2001). Assessing the attitudinal technology profiles of South African learners: A pilot study. *International Journal of Technology and Design Education* 11(1): 93–109.

Bak, H. J. (2001). Education and public attitudes toward science: Implications for the "deficit model" of education and support for science and technology. *Social Science Quarterly* 82(4): 779–795.

Bame, E. A., W. Dugger, M. de Vries, and J. McBee (1993). Pupils' attitudes toward technology-PATT-USA. *Journal of Epsilon Pi Tau* 12, no. 1:40–48.

Becker, K., and S. Maunsaiyat (2002). Thai students' attitudes and concepts of technology. *Journal of Technology Education* 13(2): 6–20.

Blomgren, R. D. (1962). An experimental study to determine the relative growth of a selected group of industrial arts education majors toward gaining an understanding of American industry. PhD diss., University of Illinois, Urbana-Champaign.

Boser, R. A., J. D. Palmer, and M. Daugherty (1998). Students' attitudes toward technology in selected technology education programs. *Journal of Technology Education* 10(1): 4–19.

Cambre, M. A., and D. Cook (1985). Computer anxiety: Definition, measurement, and correlates. *Journal of Educational Computing Research* 1(1): 37–54.

Campbell, J. (1990). *Transformations of myth through time.* New York: Harper & Row.

Committee on Information Technology Literacy (1999). *Being fluent with information technology.* Washington, DC: National Academies Press.

Conoley, J. C., and J. J. Kramer, eds. (1989). *The tenth mental measurements yearbook.* Lincoln, NE: The Buros Institute of Mental Measurements, University of Nebraska Press.

Cope, B., and M. Kalantzis (2000). *Multiliteracies: Literacy, learning and the design of social futures.* New York: Routledge.

Cuban, L., and H. Kirkpatrick (1998). Computers make kids smarter-Right? *Technos* 7(2): 26–31.

Dakers, J. R., ed. (2006). *Defining technological literacy: Towards an epistemological framework.* New York: Palgrave Macmillan.

Educational Testing Service (2002). *Digital transformations: A framework for ICT literacy.* Princeton, NJ: Educational Testing Service.

Elshof, L. (2003). Technological education, interdisciplinarity and sustainable development. *Canadian Journal of Science, Mathematics and Technology Education* 3(2): 165–184.

Eshet-Alkalai, Y. (2004). Digital literacy: A conceptual framework for survival skills in the digital area. *Journal of Educational Hypermedia and Multimedia* 13(1): 93–106.

Feng, F. (1996). The effect of gender, prior experience and learning setting on computer competency. Master's thesis, University of British Columbia.

Fisher, M. (1997). Design your future: Technology literacy competency recommendations for K-12 education. *Journal of Technology Systems* 26(1): 27–34.

Fourez, G. (1997). Scientific and technological literacy as social practice. *Social Studies of Science* 27(6): 903–936.

Frantom, C., K. Green, and E. Hoffman (2002). Measure development: The children's attitudes toward technology scale (CATS). *Journal of Educational Computing Research* 26(30): 249–263.

Garden, R. (1997). *Performance assessment in the Third International Mathematics and Science Study: New Zealand results.* Wellington, New Zealand: Research and International Section Ministry of Education.

Garmire, E., and G. Pearson, eds. (2006). *Tech tally: Approaches to assessing technological literacy.* Washington, DC: National Academies Press.

Ghiselli, E., J. Campell, and S. Zedech (1981). *Measurement theory for the behavioral sciences.* San Francisco: Freeman and Company.

Guo, R. (2006). Information and communication technology (ICT) literacy in teacher education: A case study of the University of British Columbia. PhD diss., University of British Columbia.

Hameed, A. (1988). Development of a test of technological literacy. PhD diss., Ohio State University, Columbus.

Hayden, M. (1991). The development and validation of a test of industrial technological literacy. *Journal of Technology Education* 3(1): 1–12.

Heinssen, R. K., C. R. Glass, and L. A. Knight (1987). Assessing computer anxiety: Development and validation of the Computer Anxiety Rating Scale. *Computers in Human Behavior* 3:49–59.

Hoepfl, M. (1997). Choosing qualitative research: A primer for technology education researchers. *Journal of Technology Education* 9(1): 47–63.

Howie, S. J. (1999). *Third International Mathematics and Science Study report (TIMSS-R): Executive summary.* Johannesburg: Human Sciences Research Council.

International Society for Technology in Education (2000). *National education technology standards for students.* Eugene, OR: International Society for Technology in Education.

International Technology Education Association (2000). *Standards for technological literacy: Content for the study of technology.* Reston, VA: International Society for Technology in Education.

——— (2003). *Advancing excellence in technological literacy.* Reston, VA: International Technology Education Association .

——— (2004). *Measuring progress: A guide for assessing students for technological literacy.* Reston, VA: International Technology Education Association.

Irwin, A., and B. Wynne, eds. (1996). *Misunderstanding science?* Cambridge, MA: Cambridge University Press.

Jones, M., and R. Pearson (1996). Developing an instrument to measure computer literacy. *Journal of Research on Computing in Education* 29(1): 17–28.

Kahn, R., and D. Kellner (2005). Reconstructing technoliteracy: A multiple literacies approach. *E-Learning* 2(3): 238–251.

Keirl, S. (2006). Ethical technological literacy as a democratic curriculum keystone. In *Defining technological literacy: Towards an epistemological framework,* edited by J. R. Dakers, 81–102 . New York: Palgrave Macmillan.

Kellenberger, D. (1996). Preservice teachers' perceived computer self-efficacy based on achievement and value beliefs within a motivational framework. *Journal of Research on Computing in Education* 29(2): 124–140.

Kendall, J. S., and R. J. Marzano, eds. (1997). *Content knowledge: A compendium of standards and benchmarks for K–12 education.* Eugene, OR: Mid-Continent Research for Education and Learning.

Kempton, W., J. Boster, and J. Hartley (1996). *Environmental values in American culture.* Cambridge, MA: MIT Press.

Kimbell, R. (1997). *Assessing technology: International trends in curriculum and assessment.* Buckingham: Open University Press.

Kimbell, R., K. Stables, and R. Green (1996). *Understanding practice in design and technology.* Buckingham: Open University Press.

Kimbell, R., K. Stables, T. Wheeler, A. Wosniak, and V. Kelley (1991). *The assessment of performance in design and technology.* London: Schools Examination and Assessment Authority.

Knezek, G., and R. Christensen (2000). *Instruments for assessing attitudes toward information technology.* 2nd ed. http://www.iittl.unt.edu/IITTL/publications/studies2b/ (accessed January 15, 2006).

Knezek, G. A., K. Miyashita, and T. Sakamoto (1994). *Young children's computer inventory: Final report.* http://www.tcet.unt.edu/pubs/attcomp.htm#yccifr (accessed October 15, 2005).

Kuforiji, P. O. (1992). Development and validation of an achievement test of technological literacy for senior high school students. PhD diss., West Virginia University, Morgantown.

La Porte, T., and D. Metlay (1975). Technology observed: Attitudes of a wary public. *Science* 188:121–127.

Leach, J. (1997). In search of the Holy Grail: "Scientific and technological literacy for all children, youth and adults across the world." *Studies in Science Education* 30:116–121.

Lewis, T. (2004). A turn to engineering: The continuing struggle of technology education for legitimization as a school subject. *Journal of Technology Education* 16(1): 21–39.

——— (2005). Coming to terms with engineering design as content. *Journal of Technology Education* 16(2): 37–54.

Martin, M. O., and D. L. Kelly, eds. (1996). *Third International Mathematics and Science Study technical report, volume 1: Design and development.* Boston: International Association for the Evaluation of Educational Achievement.

McLaughlin, C. (1994). Developing environmental literacy. *The Technology Teacher* 54(3): 30–34.

Miller, J. (1986). Technological literacy: Some concepts and measures. *Bulletin of Science, Technology and Society* 6:195–201.

——— (1992). *The public understanding of science and technology in the United States.* Washington, DC: National Science Foundation.

Murphy, L. L., R. A. Spies, and B. S. Plake, eds. (2006). *Tests in print VII.* Lincoln, NE: University of Nebraska Press.

Oderkirk, J. (1996). Computer literacy-a growing requirement. *Education Quarterly Review* 3(3): 9–28.

Orpwood, G., and R. Garden. (1998). *Assessing mathematics and science literacy.* Vancouver: Pacific Educational Press.

Pearson, G., and T. Young, eds. (2002). *Technically speaking: Why all Americans need to know more about technology.* Washington, DC: National Academies Press.

Pelgrum, W. J., I. A. M. Reinen Janssen, and T. Plomp, eds. (1993). *Teachers, students and computers: A cross-national perspective.* The Hague: IEA.

Petrina, S. (2000a). The politics of technological literacy. *International Journal of Technology and Design Education* 10 (2): 181–206.

——— (2000b). Review of *The Civilization of illiteracy* by Mihai Nadin. *Journal of Technology Education* 11(2): 69–70.

——— (2001). The never-to-be forgotten investigation: Luella Cole, Sidney Pressey and mental surveying in Indiana, 1917-1921. *History of Psychology* 4(3): 245–271.

——— (2003). Human rights and politically incorrect thinking versus *Technically Speaking*. *Journal of Technology Education* 14(2): 70–74.

——— (2005). How (and why) digital diploma mills (don't) work: Academic freedom, intellectual property rights and UBC's Master of Educational Technology program. *Workplace: A Journal of Academic Labor* 7(1): 38–59.

Petrina, S., F. Feng, and J. Kim (in press). Researching cognition and technology: How we learn across the lifespan. *International Journal of Technology and Design Education*.

Raat, J. H., F. de Klerk Wolters, and M. de Vries, eds. (1987). *Report PATT conference, 1987, vol. 1 proceedings*. Eindhoven, Netherlands: Eindhoven University of Technology.

Richter, J. (1980). The construction and partial validation of a scale to measure technological literacy of communication technology. PhD diss., West Virginia University, Morgantown.

Ritz, J., W. Dugger, and E. Israel, eds. (2002). *Standards for technological literacy: The role of teacher education*. New York: Glencoe/McGraw-Hill.

Robinson, M., and B. Kenny (2003). Engineering literacy in high school students. *Bulletin of Science, Technology and Society* 23(2): 95–101.

Robitaille, D., A. R. Taylor, S. Brigden, and M. Marshall (1997). *Third International Mathematics and Science Study report, volume 3: Hands-on problem-solving*. Vancouver, BC: University of British Columbia.

Rogers, G. (2005). Pre-engineering's place in technology education and its effect on technological literacy as perceived by technology educators. *Journal of Industrial Teacher Education* 42(3): 6–22.

Rose, L., and W. Dugger (2002). ITEA/Gallup poll reveals what Americans think about technology. *Technology Teacher* 61(6): 1–8.

Roth, W. M., and S. Lee (2002). Scientific literacy as collective practice. *Public Understanding of Science* 11(1): 33–56.

Saskatchewan Education. (2001). *1999 provincial learning assessment in technological literacy.* Saskatoon, SK: Saskatchewan Ministry of Education.

Selfe, C., and G. Hawisher (2004). *Literate lives in the information age: Narratives of literacy from the United States.* Mahwah, NJ: Lawrence Erlbaum.

Siciliano, P., and B. Maser (1997). *Disciplinary understanding and attitudes toward technology.* Morgantown, WV: West Virginia University.

Spies, R. A., and B. S. Plake, eds. (2005). *The sixteenth mental measurements yearbook.* Lincoln, NE: University of Nebraska Press.

Van Rensburg, S., P. Ankiewisc, and C. Myburgh (1999). Assessing South African learners' attitudes towards technology by using PATT (Pupils' Attitudes Toward Technology) questionnaire. *International Journal of Technology and Design Education* 9(2): 137–151.

Volk, K., W. M. Yip, and T. K. Lo (2003). Hong Kong pupils' attitudes toward technology: The impact of design and technology programs. *Journal of Technology Education* 15(1): 48–63.

Volk, K., and Y. M. Ming (1999). Gender and technology in Hong Kong: A study of pupils' attitudes toward technology. *International Journal of Technology and Design Education* 9(1): 57–71.

Waxman, H., M. Fen Lin, and G. M. Michko (2003). *A meta-analysis of the effectiveness of teaching and learning with technology on student outcomes.* Naperville, IL: Learning Point Associates.

Weil, M., and L. Rosen (1997). *TechnoStress: Coping with technology @work @home @play.* New York: John Wiley & Sons.

Welty, K. (1992). Technological literacy and political participation in McLean County, Illinois. *Journal of Industrial Teacher Education* 29(4): 7–22.

Woodrow, J. (1991). A comparison of four computer attitude scales. *Journal of Educational Research* 7(2): 165–187.

Wright, A. E. (1980). *Developing standards and norms for computer literacy.* Victoria, BC: British Columbia Ministry of Education.

Zuga, K. (1987). Conducting naturalistic research in technology education. *Journal of Industrial Teacher Education* 24(3): 44–58.

Assessment of Design and Technology in the U.K.: International Approaches to Assessment

Chapter 10

Richard Kimbell
Technology Education Research Unit (TERU)
Goldsmiths College, University of London

OVERVIEW

In this chapter I propose to deal with six interrelated issues:

1. The emergence of design as the core of design and technology in the U.K.
2. Understanding of design thinking
3. The learning power of design thinking
4. The challenge of performance assessment
5. An alternative approach to performance assessment: a case study
6. The close interrelationship of assessment and curriculum

THE EMERGENCE OF DESIGN PROCESSES AS THE CORE OF DESIGN AND TECHNOLOGY

Readers will have noticed that the title of this chapter gives the subject a somewhat different name: "design and technology." This is the label that was finally settled on, in the United Kingdom, after 40 years of frantic curriculum development that originally created the subject.

Design and Technology emerged from a range of craft-based traditions (e.g., woodwork, needlework, technical drawing) from about the mid-1960s. The original rationale for these craft courses was seldom made explicit, but centered on the emotional satisfaction to be achieved through "crafting" activities of various kinds. The vocational arguments that were occasionally put forward were never very convincing at a time when the world of work had already left these practices well behind.

The source of the initial transformation from craft to design and technology arose through the realization that things to be made have first to be conceived and (typically) drawn. Engaging students in the process of deciding what the object will be opened up a whole new world of ideas: the world of "design." Initially such adventures only allowed students to modify standard solutions, but gradually there evolved a realization that the object world starts life as an "ideas" world. Bringing these (object) ideas to life—to conceive new objects and to evolve new forms, structures, and purposes for them—radically transformed craft practice. By the mid-1970s, there were many U.K. schools in which design or craft and design courses became the norm. There was still plenty of crafting, but it had been reconstituted as merely the realization phase of design thinking.

At about the same time a whole strand of applied science grew up alongside this design/craft tradition, driven in part by the recognition that designed world objects are not inert. They frequently require power sources and control systems. Science teachers were unprepared for the practicalities of this world, so a new breed of technology teacher grew up alongside those in the design and craft tradition.

In 1981, midway through Margaret Thatcher's first administration, a wholesale reappraisal of the curriculum was undertaken in which the various subjects of the curriculum were called upon to justify their existence. For the first time, government wanted some unequivocal statements about what was being taught in schools, why, and how we know what youngsters can do as a result. What is math, and why do we study it? Similar questions were asked about history, music, design, and the rest. The group of subjects that clustered around our area of the curriculum had to settle upon a single coherent title, and the best that could be agreed upon was Craft Design and Technology (CDT).

Centralizing control over the curriculum had begun, ironically, by Thatcher, and resulted a decade later in the establishment (in 1990) of the national curriculum. ("Ironically" because in all areas of political life, Thatcher was a de-centralizer, celebrating the power of the open market. Nevertheless, with education, she initiated processes of enormous centralization and government control, both of curriculum and of assessment.) By then the title of the subject had evolved again—to Design and Technology—reflecting a further diminution in craft activity and celebrating the centrality of design thinking and the critical role of technology in bringing that thinking to life:

> ... whereas most, but not all design activities will generally include technology and most technology activities will include design, there is not always total correspondence. Our use of design & technology as a unitary concept ... does not therefore embody redundancy. It is intended to emphasize the intimate connection between the two activities as well as to imply a concept which is broader than either design or technology individually, and the whole of which we believe is educationally important. (Department of Education and Science/Welsh Office 1988, ¶ 1.5–1.6)

The British curriculum has been a slow-moving beast, as Williams has observed: "The fact about our present curriculum is that it was essentially created by the 19th Century, following some 18th Century models and retaining elements of the mediaeval curriculum near its centre" (Williams, 1965). Design and technology, however, has been a dramatic exception to that rule, emerging through a series of guises to its current formulation, and all within 40 years, or one professional lifetime. And at the heart of the new national curriculum formulation is a vision statement that stands as a powerful statement of learning purpose:

> Design and technology prepares pupils to participate in tomorrow's rapidly changing technologies. They learn to think and intervene creatively to improve the quality of life. The subject calls for pupils to become autonomous and creative problem solvers, as individuals and members of a team. They must look for needs, wants and opportunities and respond to them by developing a range of ideas and making products and systems. They combine practical skills with an understanding of aesthetics, social and environmental issues, function and industrial practices. As they do so, they reflect on and evaluate present and past design and technology, its uses and effects. Through design & technology, all pupils can become discriminating and informed users of products, and become innovators. (Department for Education and Employment/ Qualifications and Curriculum Authority [DfEE/QCA], 1999, p. 15).

In operationalizing this vision, students are taught to observe contexts within which they might identify and pin down a task. They then investigate the factors bearing upon it and seek to generate ideas that might lead to a solution. These product concepts are then developed through models and working drawings and a prototype of the product is built. This is then reviewed in context, with the intended user, to see how well it serves the original purpose. Some examples of major projects (age 16–18) are illustrated in figures 1 through 3.

Figure 1. An automatic fish-feeder designed to feed a known amount of food into the tank every day for 18 days, so the student can take a two-week holiday.

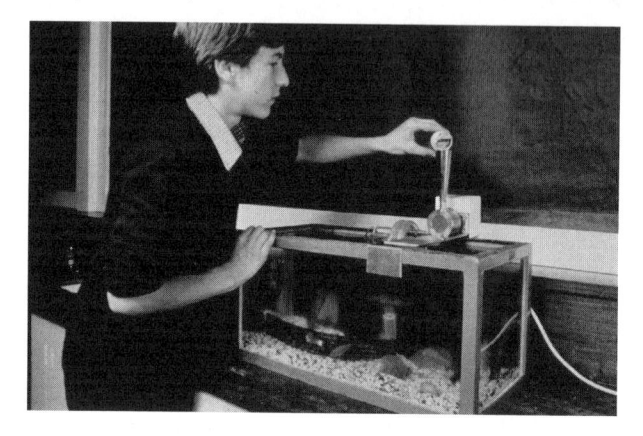

Figure 2. A one-handed can opener designed for use by a grandparent disabled by arthritis in one hand.

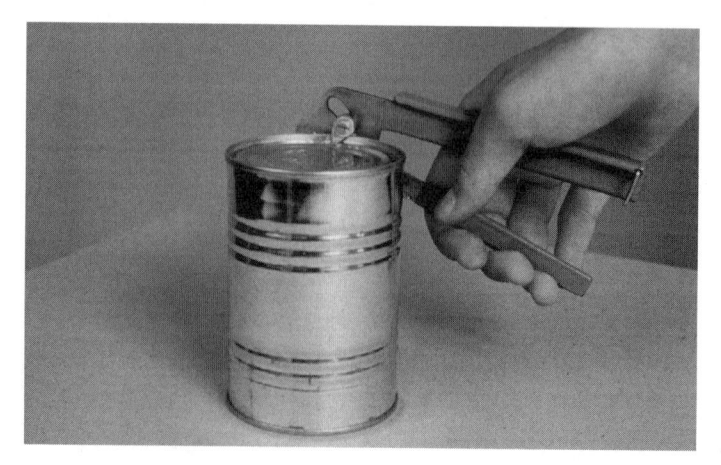

Figure 3. A nursery school chair that is three chairs. Using its three positions it can "grow" from a toddler chair to a junior chair. It also has built-in storage.

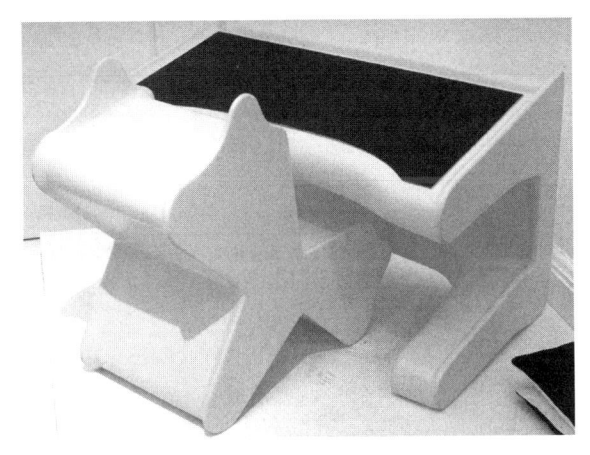

UNDERSTANDING OF DESIGN THINKING

During the late 1980s, the Department of Education and Science (DES) commissioned a research project to establish the national levels of capability in design and technology among 15-year-old students across the U.K. The project was managed by a research branch of the DES called the Assessment of Performance Unit (APU). I directed that project, and one of our first self-imposed tasks was to seek understanding of the cognitive/conceptual processes that enable humans to do the kinds of things that are demonstrated through the projects outlined above. In the light of literature review, and analysis of assessment activities with approximately 10,000 students in 700 schools across the U.K., we came to the following conclusion, which is illustrated in Figure 4:

> When engaged in a task, ideas are inevitably hazy if they remain forever in the mind, and this inhibits their further development. By dragging them out into the light of day as sketches, notes, models or the spoken word, we not only encourage the originator to sharpen up areas of uncertainty, but we also lay them open to public scrutiny.... The act of expression pushes ideas forward ... and the additional clarity that this throws on the idea enables the originator to think more deeply about it, which further extends the possibilities of the idea.... Concrete expression (by whatever

means) is therefore not merely something that allows us to see the designer's ideas, it is something without which the designer is unable to be clear what the ideas are. (Kimbell, Stables, Wheeler, Wosniak, and Kelly, 1991, p. 20)

While this work for the APU was based on our experience of novice designers and technologists at school, an exactly parallel finding emerged at about the same time from a study of expert technologists. Gorman and Carlson (1990) at the University of Virginia used the entire Edison and Bell archive of papers, sketches, notes, and experiments to create an account of the process through which these inventors had developed the telephone. There was no connection between this study in the USA and our own in the U.K., but the conclusion they arrived at was astonishingly parallel to our own:

> ... the innovation process is much better characterized as a recursive activity in which inventors move back and forth between ideas and objects. Inventors may start with one mental model and modify it after experimentation with different mechanical representations, or they may start out with several mechanical representations and gradually shape a mental model. In both cases, the essence of invention seems to be the dynamic interplay of mental models with mechanical representations. (Gorman and Carlson, 1990, p. 55)

We used the term "iterative" to describe the interaction of the internal (mind's eye) images and their expression in drawings and models. Gorman and Carlson use the term "recursive," but the point remains the same.

Figure 4. A schematic illustration of the design process.

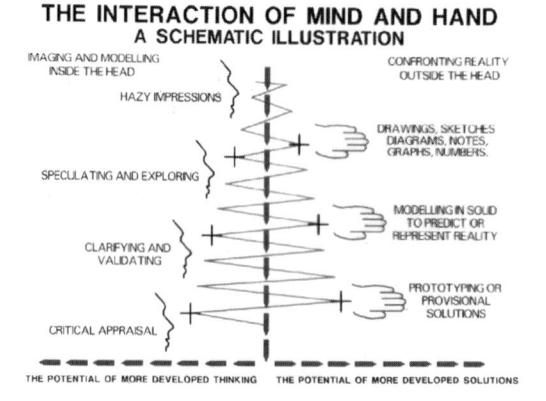

THE INTERACTION OF MIND AND HAND
A SCHEMATIC ILLUSTRATION

IMAGING AND MODELLING INSIDE THE HEAD — CONFRONTING REALITY OUTSIDE THE HEAD

HAZY IMPRESSIONS

DRAWINGS, SKETCHES, DIAGRAMS, NOTES, GRAPHS, NUMBERS.

SPECULATING AND EXPLORING

MODELLING IN SOLID TO PREDICT OR REPRESENT REALITY

CLARIFYING AND VALIDATING

PROTOTYPING OR PROVISIONAL SOLUTIONS

CRITICAL APPRAISAL

THE POTENTIAL OF MORE DEVELOPED THINKING   THE POTENTIAL OF MORE DEVELOPED SOLUTIONS

We would therefore argue that the core of designer-like thinking processes lies in the interplay of internal imaging in the "mind's eye" (Kimbell et al., 1991, p. 21) and externalizations of these images to test them against reality and hence to refine and extend them.

> Cognitive modeling by itself—manipulating ideas purely in the mind's eye—has severe limitations when it comes to complex ideas and patterns. It is through externalized modeling techniques that such complex ideas can be expressed and clarified.... It is our contention that this inter-relationship between modeling ideas in the mind and modeling ideas in reality is the cornerstone of capability in design and technology. It is best described as "thought in action." (p. 21)

We used the expression "modeling ideas in the mind," which more technically might be described as cognitive modeling:

> The conduct of design activity is made possible by the existence in man of a distinctive capacity of mind ... the capacity for cognitive modeling.... [The designer] forms images "in the mind's eye" of things and systems as they are, or as they might be. Its strength is that light can be shed on intractable problems by transforming them into terms of all sorts of schemata ... such as drawings, diagrams, mock-ups, prototypes and of course, where appropriate, language and notation. These externalizations capture and make communicable the concepts modeled. (Archer and Roberts, 1992, p. 15)

THE LEARNING POWER OF DESIGN THINKING

When the broad process of design thinking described here is interpreted into specific kinds of capability, a number of qualities rise to the surface. These are presented below, and have been elaborated by reference to the design literature.

Unpacking Wicked Tasks

Designing involves creating potential solutions to "wicked" problems. This is much more complex than typical problem solving. Buchanan (1996) points out that the problem for designers is to conceive and plan what does not yet exist, and this creative projection into the future is only very inadequately described as problem solving. The student's task is seen in terms of a wicked design problem, which

- is individual (i.e., each is unique).
- has no definite formation.
- has no stopping rules (i.e., development can just go on and on).
- cannot be true or false (but can be better or worse).
- has no complete list of operations.
- is capable of multiple solutions.
- has no definitive "truth" test. (Buchanan, 1996)

The design literature is full of accounts of the complexity of design tasks that have no right answers or set procedures. The learning challenge is to bring clarity to this chaos: to unpack this messiness to clarify what (and who) is involved. Moreover, this process is not merely a starting strategy, but is rather a progressive one, so the designer is continually and progressively unpacking the task to identify its constituent elements and their significance.

"Playing" with Reality

There is an expanding body of evidence that links designing to ideas of playfulness. Kathy Sylva, for example, has researched extensively the value of structured play with very young children (Sylva et al., 1976), and Papanek (1995) talks of design as "goal-directed play" (p. 7). The significance of this playfulness is that it allows imagination to operate without too tight a framework of constraints.

> The concept of "what might be"—being able to move in perception and thought away from the concrete given of "what is" to "what was, what could have been, what one could try for, what might happen" and ultimately to the purest realms of fantasy—is a touchstone of that miracle of human experience, the imagination. (Singer and Singer, 1990, p. 241)

To engage in designing processes requires us to be openly speculative, considering multiple "what-if" possibilities without automatically ruling them out as impractical or silly. Often the weirdest ideas emerge as fantastic design solutions, like the "beam of light" that Foster imagined as the new millennium bridge across the Thames in London. It was some very clever engineering that made the idea work (when they finally stopped it from wobbling), but it was the playful idea of a beam-of-light bridge that makes it so stunning. Stables (1992) concluded that one of the key qualities that young children bring to designing is this level of imagination, which at critical moments needs to be challenged by a dose of reality: "This ability to handle fantasy and reality simultaneously could be argued to be of fundamental importance to the design and technologist. To be able to conceive ideas that push out the boundaries of the possible, while mediating these ideas through a grip on reality" (p. 111).

Thinking as Someone Else

Design is about improvement, and the concept of improvement is essentially value-laden. Good design practice therefore seeks to identify the stakeholders in any task and to make their values explicit from the outset. In a recent research project in schools we invited students, teachers, janitors, parents, principals, and the local community police and fire officers to speculate on how we might improve the safety and security of the school. Inevitably in the isolation of their own groupings they all saw the task differently (through their own eyes and with their own priorities); janitors, for example, wanted more locks and a stricter regime of locking doors and windows. When we put them together to discuss their ideas, the very different notions of "security" became apparent. Fire officers, for example, were horrified about the danger of all the locked doors. Designers need to be able to see the task through others' eyes—getting inside the values and priorities of all the stakeholders. Forcing these values to become explicit is a critical step.

Modeling Possible Futures

Part of the problem of dealing with the new is the fact that it is very difficult to make the necessary judgments about it if we cannot first create a realistic simulation of what it is going to be like. Design students, therefore, must continually model their concepts of the future. The value of modeling relates to the issue of risk in the new and innovative, since modeling exposes the "consequences" of decision-making. Risks can, and invariably need to be, taken in the thinking and development that eventually emerge as the outcome. But, processes of modeling allow the designer to mitigate and offset the risk by testing out its consequences "before" coming to a conclusion.

Managing Complexity

The wickedness of design tasks places complex management demands on students: managing (and optimizing) time, cost, materials, production processes, technical performance, and much more in ways that enable them to complete their task. At the end they typically have to bring together all trains of thought and development into a single, holistic solution. They need to be integrative thinkers while managing the messy and often contradictory strands of thought within a project.

Task-Related Knowledge

Since the demands in any task may vary considerably, students need to develop robust, self-confident strategies to inform their designing by acquiring appropriate resources of knowledge and skill. This is widely acknowledged even in the examination rubrics that dominate school-based assessment in design and technology:

> When embarking upon a new design, the package of knowledge and skills necessary for the success of the venture will emerge as the design progresses, and so the need to acquire knowledge and skills (and sometimes extend the boundary of knowledge and devise new skills) becomes a clear requirement for the designer. (Threlfall, 1980, p. 7)

As design and technology was becoming formulated in the early 1980s, the Department of Education and Science produced its booklet *Understanding Design & Technology*, in which the very same view of knowledge was adopted:

> The designer does not need to know all about everything so much as to know what to find out, what form the knowledge should take, and what depth of knowledge is

required for a particular purpose. (Department of Education and Science, 1981, p. 12)

To summarize, this is neither an exhaustive list of the learning challenges that are explicit in design and technology, nor is it a list that applies exclusively to design and technology, since they are in reality generic cognitive skills. But although modeling ideas has meaning in mathematics and science, it has highly explicit meaning in design and technology. While optimizing values and thinking as someone else has meaning in social studies or history, it is at the sharp end of design and technology where decisions made by students will condition whether the solution works or not for those others. The collected list of qualities identified here amounts to a manifesto for design and technology as a learning activity.

There is an additional element that dramatically enhances the learning power of design and technology. With all the qualities listed above, students in design and technology not only have to do those things (e.g., model ideas, manage complexity, think as someone else), they must become consciously aware of the fact that they have to do those things. Abstract cognitive processes have been brought to life, have been embodied in material form, and have been presented in ways that unavoidably make the processes explicit to students. As they become aware of their own cognitive processes—become capable meta-cognitive designers—they reach the ultimate reward. I would argue that the qualities listed above are more evident in design and technology than they are in other subjects, and they are also more developed through design and technology than through other subjects.

THE CHALLENGE OF PERFORMANCE ASSESSMENT

In assessment terms, a number of things follow from the nature of the design and technology description that has been offered above. First, and perhaps most critically, we have described design and technology as a "process" rather than as a body of knowledge or skills. Students do not learn "about" design and technology, they learn to "become" designers and technologists. If design and technology is concerned with enabling students to bring about purposeful change in the made world, then our assessments must be focused on students' abilities to operationalize that process. How well can they do it? Are there parts of the process that they find harder than others? This, in terms of design and technology, is performance assessment, and in a more generalized form would be described as "authentic" assessment:

> Assessment is authentic when we directly examine student performance on worthy intellectual tasks. Traditional assessment, by contrast, relies on indirect or proxy "items"... from which we think valid inferences can be made about the student's performance at those valued challenges. (Wiggins, 1990, ¶1)

More than this, however, we have to acknowledge what is perhaps the most difficult consequence for those involved in assessment. Design challenges may be focused on anything (e.g., designing a seating system, a heating system, a messaging system) and may appropriately result in products of radically different kinds (e.g., a necklace, a lawnmower, or a sailing dinghy). In other words, "the subject matter of design is potentially universal in scope, because design thinking may be applied to any area of human experience" (Buchanan, 1996). Buchanan's point is that each new design task involves its own individually constructed subject matter, and this has led thinking in the U.K. to acknowledge assessments that do not predefine the knowledge content:

> It is ... much more difficult to be precise about the areas of knowledge that are essential, because it is just not possible to define exactly what one will be required to know about in advance of the activity. As a design exercise develops it *may* become necessary to know about the manner in which arthritis progresses through old people's joints ... or it *may* be necessary to know how animals behave in confinement.... It must be accepted that [pupils] will have the ability to seek out the specific knowledge that they require for any exercise. (Secondary Examinations Council, 1986, pp. 14–15)

Buchanan extends this to become a philosophical argument, asserting that design is fundamentally concerned with the "particular," and that there is no science of the particular. This, he argues, has led to an interesting consequence:

> We have been slow to recognize the peculiar indeterminacy of subject matter in design and its impact on the nature of design thinking. As a consequence each of the sciences that have come into contact with design has tended to regard design as an "applied" version of its own knowledge, methods, and principles.... We have the odd recurring situation in which design is alternately regarded as "applied" natural science, "applied" social science, or "applied" fine art. (Buchanan, 1996, p. 18)

The cutting edge of this argument for assessment purposes is that we should be less concerned with assessing what students "know," and more with what they "can do." Further, it argues that rather than worry about testing what students know, we should rather be testing their ability to find out what they "need" to know in order to facilitate their designing.

This has continually forced our assessment thinking to the conclusion that the only valid way to assess students' ability in design and technology is to set them a task and see what they can do with it. Do they have robust, creative processes that enable them to perform? Project-based assessment has therefore been at the heart of the evolving story of design and technology, and all the projects illustrated in figures 1 through 3 of this chapter were undertaken by students as assessment projects at age 16 for the General Certificate of Secondary Education (basic school-leaving assessments) or at age 18 for Advanced level (university entrance).

The tradition of these assessment projects is that students identify a task; develop initial concept designs; explore these through a range of imaging and modeling techniques; try out and negotiate the emerging solution with its user; finalize the design; and manufacture a prototype. It is very uncommon for practice to go beyond the one-off, prototype stage, although it does happen. The project at age 16 may take up to 30 hours of curriculum time and many hours beyond that in working on the design portfolio. At age 18 it could be significantly longer.

The assessment at the end of the project will typically involve an exhibition constructed by students to display their work. It will include a detailed design portfolio, or design sketchbook, involving a discussion of the task, a series of evolutionary models, results derived from user testing, final drawings, and a working prototype. Ideally, this will be taken back to the intended user for full evaluation, and may also involve market potential analysis, including comments from manufacturers of related products.

Each year in the U.K. approximately half a million students are assessed in this way for design and technology awards at GCSE or A level. And some of this work is truly astonishing. Two years ago—after a major train disaster resulting from a cracked rail—a student developed a system for testing rails that proved superior to anything in use in the industry. He is now not only at a university studying engineering but is also wealthy, having been employed as an independent consultant by Railtrack, the railway infrastructure company. At a simpler level, two 16-year-olds redesigned a windshield washer system so that it automatically refilled the washer bottle from rainwater filtered from the windshield. Their solution was bought by Smiths Industries, an electrical products manufacturer.

The rationale for design and technology does not, of course, require such commercial success, for the real benefit of the experience is the "learning benefit," made explicit in the national curriculum statement, which states:

> Design and technology prepares pupils ...
> - to intervene creatively to improve the quality of life.
> - to become autonomous and creative problem solvers.
> - to combine practical skills with an understanding of aesthetics, social, and environmental issues and industrial practices.
> - to reflect on and evaluate present and past design and technology, its uses, and effects.
>
> Through design and technology, all pupils can become discriminating and informed users, and can become innovators. (DfEE/QCA, 1999, p. 7)

It is important to acknowledge, however, a number of problems that are almost (but not quite) inevitably associated with this authentic vision of performance assessment. The literature argues that students' performances will vary from task to task, so to take one particular task as indicative of the level achievable by a student may be unfair. Also, at the practical level of the school workshop/design studio, there is little other than teachers' professional ethics to ensure that they do not unreasonably support an individual's work. Moreover, since the design portfolio is something that students carry with them throughout the months of the project, there is little to prevent the influence of parental support. All of these factors tend to raise questions about the reliability of the data resulting from such performance assessments. The high level of face validity for such authentic assessment activities can appear to be compromised by questionable control issues.

These issues have received focused attention at the Qualifications and Curriculum Authority (QCA), the regulatory body responsible for all school assessment matters in the U.K., and a number of strategies have evolved in an attempt to resolve them. These include assessments taken at multiple occasions, inspections of schools and their systems, double-blind marking of assessment portfolios, and codes of conduct that schools must sign on to (including issues to do with parental influence).

None of these will absolutely assure that project-based assessment will yield data that is as reliable as timed, blind, written papers undertaken in the anonymity of the examination hall. But QCA and the Design & Technology Teachers' Professional Association (DATA) would be very reluctant to take such reliability if it meant trading away the authentic validity of the current assessment regime. Having seen assessment practices evolve over the last 30 years, the vast majority of students, teachers, and schools believe that the current arrangements are appropriate.

CASE STUDY: AN ALTERNATIVE APPROACH TO PERFORMANCE ASSESSMENT

Nonetheless, QCA, in association with DATA, has begun to look at ways to supplement the conventional assessment process. Specifically, in 2002 they commissioned Technology Education Research Unit (TERU) at Goldsmiths College to explore and develop an approach to performance assessment that might enable some of the weaknesses of project-work assessment to be countered.

Through a series of trials in schools throughout the country, the format of the assessment evolved into a six-hour task, using two consecutive mornings of three hours. In that time, students start with a task and work from an initial concept to the development of a prototype solution—a "proving model" to show that their ideas will work. The whole six hours is run with an administrator script that choreographs the activity through a series of subtasks, each of which is designed to promote evidence of students' thinking in relation to their ideas. These steps in the process all operate in designated spaces in a response booklet.

The structure shown in Figure 5 is characteristic of the activities so far developed. The task (called Light Fantastic) centers on redesign of a light bulb packaging box so that once the bulb is removed, the package can be transformed into a lighting feature, either by itself or in association with other empty light-bulb packaging.

In the example shown in Figure 5, the light-bulb box has been redesigned as a tapering pentagon that fits with many more to build into a complete hanging sphere (or hemisphere standing on a surface). With a bulb suspended at the center of the sphere, the letter cutouts (with inset lighting film) project the letters onto the wall. The student titled the prototype "Your Name in Lights."

In addition to the models themselves, students' workbooks (folding A2 sheets) capture their evolving ideas over the six hours and their reflections on them. They also capture the contributions of students' teammates and, running down the spines, are the photo records of the evolving 3-D outcomes (Figure 6).

Figure 5. Student-developed prototype for lighting design task.

This assessment format is specifically intended to encourage design innovation, but in the process it also irons out a number of conventional problems with project assessment. Specifically, it levels the playing field in terms of time, facilities, teacher role, parental input, task type, and teamwork. At the same time it enables very different kinds of responses to

Kimbell

flourish. Some students prefer to evolve their ideas more through modeling while others prefer to draw. The assessment trials suggest that assessors—using double-blind marking—can make reliable judgments about the resultant work.

Figure 6. Sample student workbook page for lighting design task.

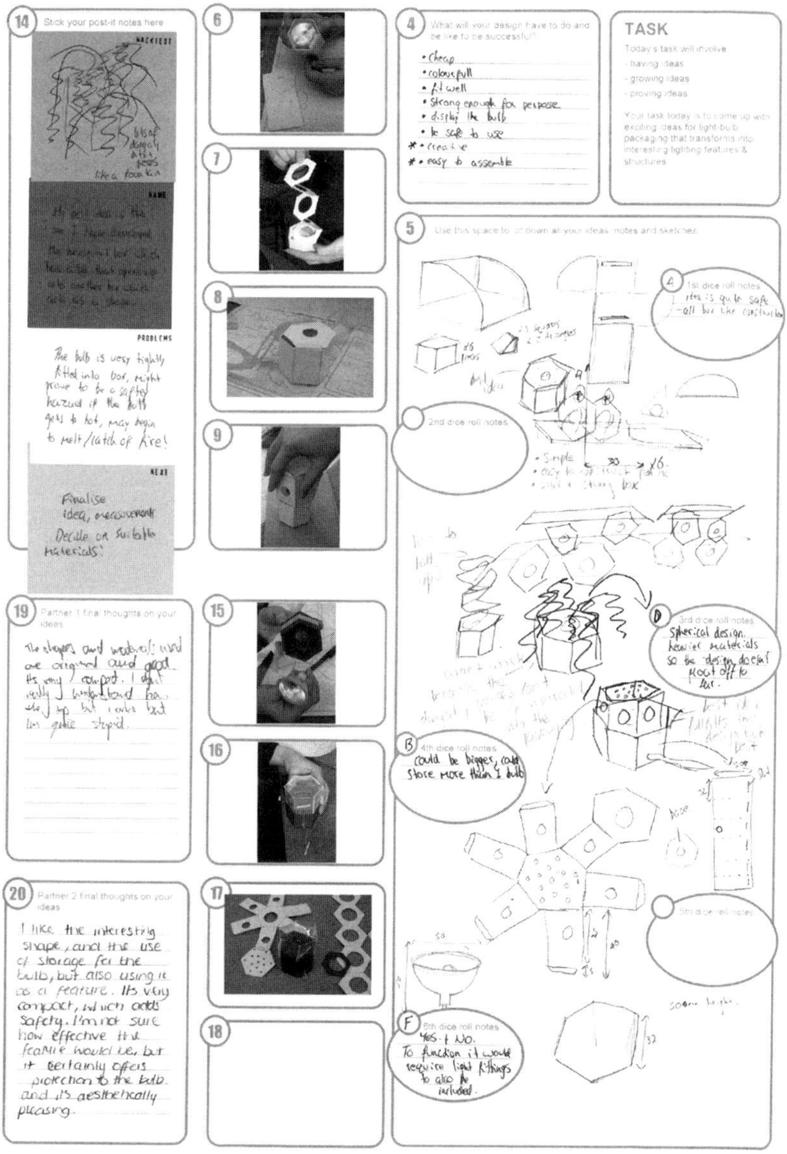

= 197 =

Furthermore, having developed and tested eight different tasks, all of which fit into the same activity framework and hence the same response booklet format, we were able to explore the context effect on performance across tasks. It is much smaller than the literature might suggest, probably because of the common framework being used (along with the standardized administrator script) for structuring the activity and making the assessments.

On a 12-point holistic scale across the whole sample, the mean difference between test 1 and test 2 was +/- 1.6. On average, students scoring 9 on test 1 would score between 7.4 and 10.6 on test 2. Those scoring 3 on test 1 would score between 1.4 and 4.6 on test 2. This indicates a level of consistent discrimination in the range of performance of individual students. In 80% of the cases, the scores on two tests varied by 2 marks or less on a 12-point scale (Kimbell et al., 2004).

The research findings were reported to DfES/QCA in December 2004, and since that time, the approach has been adopted for national assessment purposes by one of the GCSE examining bodies. Our approach operates as a kind of hybrid: it is neither an open-ended piece of project work nor a closed examination. It remains faithful to the concept of performance assessment, and judging by the reaction in schools, it is seen by teachers as a positive development thus far.

THE CLOSE INTERRELATIONSHIP OF ASSESSMENT AND CURRICULUM

It is realistic to claim that design and technology in the school curriculum has been shaped by the priorities of assessment. The two stories (defining the discipline and assessing pupil capability) are intimately interwoven, and it is easy to see why.

In the 1970s and 1980s, design and technology was a newly emerging phenomenon and it did not arrive through a single coherent birth process, but rather from a messy series of initiatives that contributed to parts of the story. In 1980, before Mrs. Thatcher's era of forced consolidation, there were approximately 120 different syllabi descriptions for subjects associated with what we now call design and technology. Some accentuated graphics, some technology, some crafts, some electronics, some pneumatics, some photography, and so on. They each assessed their own practices in various ways. As design and technology began to coalesce, we created new kinds of assessments that sought to capture the qualities of the evolving discipline, and these assessments provided a very direct feedback mechanism into the curriculum. They began to show us through the nature of their qualities that it was possible to develop this evolving beast.

As these qualities began to be made explicit through assessment processes, the curriculum statements reinforced them, effectively seeking out higher levels of performance in students. This period of 20 years, from about 1970 to 1990, gives life to the old adage that we assess what we "can" assess rather than what we "should" assess. It is true that we were experimenting with what we could assess, but in that process we illuminated a series of qualities that proved to be very important in the emergence of design and technology. Admittedly, however, it was a painful process.

> At every turn those of us working within the field of technology were forced to define, and redefine, and re-redefine what we were doing. And in most cases this pursuit of ever-tighter definition was motivated by the needs of assessment: For GCE, for CSE, for National Criteria, for GCSE, for A level, for national curriculum. (Kimbell, 1997, p. 5)

Despite the pain involved, it is possible to argue (in fact I do believe) that the processes that created design and technology were enriched by the contribution of assessment. The process began in the 1960s, and the highly autonomous nature of schools and curriculum at that time enabled teachers to do more or less whatever they thought would be interesting for students. But, being responsible, they sought ways to attach value to their activities by awarding assessment grades. By the end of the 1970s we had a highly divergent curriculum that contained most of the building blocks of what is now design and technology. The Thatcher decade of consolidation (1980–90) then forced us to refine, and refine again, the qualities that we believed to be central to our collective goals.

Creative endeavor in any field will necessarily involve all of these forces: open-ended exploration, hard-nosed trade-offs, and consolidation. Design and technology stands as a classic example of how this process has worked in the U.K. curriculum.

The role of assessment in this process has been to make very clear to us how these curriculum experiences emerge as student capabilities. There has been just as much experimentation with assessment practices as there has with curriculum, and the forms of assessment now are very different from that which was commonplace in the 1960s or 1970s. We have emerged with a national curriculum statement that is dominated by procedural learning, that sets student capability at the top of the agenda, and that is underpinned by an assessment regime that is premised on performance.

DISCUSSION QUESTIONS

1. In your opinion, what would be the ideal set of assessment processes for your curriculum?
2. What are the forces that shape your curriculum and the assessment processes you use? Which of these forces are compatible with your goals for the curriculum?
3. What categories of students are particularly empowered by your curriculum, and which are particularly rewarded by your assessment processes? (Note that these may be different.)
4. Why do students choose to take part in your curriculum, and what qualities do you think they value? Are these values part of the assessment process?
5. Assessing students' capability with processes is more complex than assessing their ability to retain content knowledge. So why bother?

REFERENCES

Archer, B., and P. Roberts (1992). *Design and technological awareness in education*. In Modelling: The language of design (Design Occasional Paper No. 1), edited by P. Roberts, B. Archer, and K. Baynes. Loughborough, England: Loughborough University of Technology, Department of Design & Technology.

Buchanan, R. (1996). *Wicked problems and design thinking*. In The idea of design: A design issues reader, edited by V. Margolin and R. Buchanan, 3–20. Cambridge, MA: MIT Press.

Department of Education and Science (DES) (1981). *Understanding design and technology.* London: HMSO.

Department of Education and Science/Welsh Office (DES/WO) (1988). *National curriculum: Design & technology working group: Interim report.* London: DES.

Department for Education & Employment/Qualifications and Curriculum Authority (DfEE/QCA) (1999). *The national curriculum for England.* London: HMSO.

Gorman, M. E., and W. B. Carlson (1990). Interpreting invention as a cognitive process: The case of Alexander Graham Bell, Thomas Edison, and the telephone. *Science, Technology, and Human Values* 15(2): 131–164.

Kimbell, R., K. Stables, T. Wheeler, A. Wosniak, and V. Kelly (1991) *The assessment of performance in design & technology: The final report of the APU project.* London: School Examinations and Assessment Council/Central Office of Information for HMSO.

Kimbell, R. (1997). *Assessing technology: International trends in curriculum and assessment.* Buckingham, UK: Open University Press.

Kimbell, R., S. Miller, J. Bain, R. Wright, T. Wheeler, and K. Stables (2004). *Assessing design innovation: A research and development project for the Department of Education & Skills (DfES) and the Qualifications and Curriculum Authority* (QCA). London: Goldsmiths College.

Papanek, V. (1995). *The green imperative: Ecology and ethics in design and architecture.* London: Thames and Hudson.

Secondary Examinations Council (SEC) (1986). *Craft, design & technology GCSE: A guide for teachers.* Buckingham, UK: Open University Press.

Singer, D., and J. Singer (1990). *The house of make believe.* Cambridge, MA: Harvard University Press.

Stables, K. (1992). The role of fantasy in contextualizing and resourcing design and technological activity. In *IDATER 92: International conference on design and technology educational research and curriculum development,* edited by J. Smith, 110–115. Loughborough: Loughborough University of Technology.

Sylva, K., J. S. Bruner, and P. Genova (1976). *The role of play in the problem-solving of children 3–5 years old. In Play, its role in development and evolution,* edited by J. Bruner, A. Jolly, and K. Sylva, 244–257. London: Penguin Books.

Threlfall, P.M. (1985). "A" level design & technology: The identification of a core syllabus. (A report prepared for the Council for National Academic Awards (CNAA) and Standing Conference on University Entrance (SCUE) Working Group.)

Wiggins, G. (1990). *The case for authentic assessment. Practical assessment, research, and evaluation,* 2(2). http://PAREonline.net/getvn.asp?v=2&n=2 (accessed September 12, 2006).

Williams, R. (1965). *The long revolution.* London: Penguin Books.

Assessment in Science and Mathematics: Lessons Learned

Chapter 11

Maria Araceli Ruiz-Primo
University of Colorado at Denver/Stanford Education Assessment Laboratory

INTRODUCTION

What constitutes technological literacy achievement? How should we assess whether a person is technologically literate or whether a student has achieved technological literacy? The same type of question is asked in the domains of science and mathematics. Professional organizations have proposed some answers by publishing standards, guidelines, and blueprints, and these questions are in some way addressed in every assessment framework. However, controversies have always emerged, and a consensus on what students should know and be able to do, what to assess, and how to assess it has never been reached.

The lack of a theory underlying the assessment of school achievement and the growth of competence may be the main reason for this conceptual disarray (Glaser and Silver, 1994; Snow, 1993). We have to develop a theory of subject-matter achievement assessment, or a psychology of test design (using Snow's terms), that combines the cognitive, measurement, and discipline perspectives in designing assessments for a particular content domain (Pellegrino et al., 2001). It can be argued that preferences for one or another type of assessment and its characteristics should rest on what we know about cognitive processes, educational measurement, and the characteristics of the content domain to be assessed.

In this chapter, I present some ideas that attempt to provide the first steps toward conceptualizing achievement and assessment. The ideas are organized around three aspects of achievement assessment: (a) a conceptual framework on achievement, (b) a framework for developing and evaluating assessments, and (c) how these frameworks relate to the form and function of assessment. The assessment aspects discussed have been developed and implemented in the context of science education. Therefore, examples used are mainly in the science domain.

First, I present a framework on achievement. The framework conceptualizes achievement as a multidimensional construct with direct implications for developing and/or interpreting students' performance. The framework identifies four types of knowledge that are characteristic of competency or achievement in science (Li, 2001; Li, Ruiz-Primo, and Shavelson, 2006; Shavelson and Ruiz-Primo, 1999; Ruiz-Primo, 1997, 2002): declarative knowledge, or *knowing that;* procedural knowledge, or *knowing how;* schematic knowledge, or *knowing why;* and strategic knowledge, or *knowing when, where, and how knowledge applies.* Next, I present a strategy my colleagues and I have worked on for developing and evaluating assessments. The strategy involves different types of analyses—conceptual, logical, and empirical. Finally, I discuss *function* and *form* in student assessment. I focus on the distinction between externally mandated assessment and classroom assessment, and their implications for designing, implementing, and evaluating assessments.

TOWARD A DEFINITION OF ACHIEVEMENT FOR ASSESSMENT PURPOSES

Educational achievement should refer to what students *know and are able to do* in a particular domain after instruction. It does not refer merely "to the amount of knowledge accumulated but its organization or structure as a functional system for productive thinking, problem solving, and creative invention in the subject area as well as for further learning" (Messick, 1984, p. 156). As a result of effective learning and educational experiences, then, students should develop a knowledge that is structured in such a way that it contributes to their ability to think and reason with what they know (Baxter et al., 1996; Glaser and Silver, 1994). Students should not be granted achievement in a domain unless their performance is consistently correct and unless they succeed on a wide variety of tasks that are reasonably defined as relevant to that domain (Gelman and Greeno, 1989).

Any notion of academic achievement should consider at least two dimensions: a *content* dimension (the subject matter dealt with in a field of study), and a *cognitive* dimension (the intellectual processes that we would like students to engage with the content at hand). The content dimension is usually organized in domains (such as life science, chemistry, physics, earth science, algebra, geometry, and technology) in both standards and assessment frameworks. Examples include the Third International Mathematics and Science Study (TIMSS, 2003) and the Organisation for Economic Co-operation and Development's (2003) Programme for International Student Assessment (PISA). The cognitive dimension is also organized around domains that represent cognitive processes. Domains also vary by content. For example, TIMSS (2003) proposed three cognitive domains in science (factual knowledge, conceptual understanding, reasoning and analysis) and four in mathematics (knowing facts and procedures, using concepts, solving routine problems, and reasoning). Similar differences are observed in PISA.

The framework for conceptualizing achievement presented in this section focuses on the cognitive dimension. The framework, as it has evolved, has been described in other documents (Li, 2001; Li, Ruiz-Primo, and Shavelson, 2006; Shavelson and Ruiz-Primo, 1999; Ruiz-Primo, 1997, 2002; Ruiz-Primo, Shavelson, Hamilton, and Klein, 2002). It draws upon a body of related research (Alexander and Judy, 1988; Anderson, 1997; Bybee, 1996; Chi, Feltovich, and Glaser, 1981; Chi, Glaser, and Farr, 1988; de Jong and Ferguson-Hessler, 1996; Pellegrino et al., 2001; Sugrue, 1995; White, 1999), and it has been empirically tested in the context of science assessments, with confirming results around the categories proposed (Li, 2001; Li and Shavelson, 2001; Li, Ruiz-Primo, and Shavelson, 2006; Yin, 2005).

The framework is based on four interdependent types of knowledge. It focuses on the types of knowledge that result from learning within a content domain (either by formal or informal schooling, or by self-study) and not on the types of knowledge that are universally acquired in the course of normal development (Pellegrino et al., 2001). It rests on the idea that expertise is necessarily constrained to a subject matter or content domain (Chi, Feltovich, and Glaser, 1981). Evidence shows that an expert's knowledge and skills in one domain are not transferable to another. Another underlying assumption is that the types of knowledge proposed reflect, to a certain degree, the nature of subject-matter expertise. That is, types of knowledge lie in a continuum from concrete to abstract, from bits of information to high levels of organized knowledge. Therefore, it should be assumed that higher levels of achievement are linked to certain types of knowledge.

Types of Knowledge

Declarative Knowledge (Knowing That)

Declarative knowledge includes knowledge that ranges from discrete and isolated content elements, such as terminology, facts, or specific details (e.g., vocabulary such as mass or density), to more organized knowledge forms, such as statements, definitions, knowledge of classifications, categories, principles, and theories (e.g., mass is a property of an object; density is expressed as the number of grams in 1 cubic centimeter of the object).

Procedural Knowledge (Knowing How)

Procedural knowledge involves knowledge of skills, algorithms, techniques, and methods. Usually, it takes the form of "if-then" production rules or a sequence of steps (Anderson, 1983). It ranges from motor procedures (e.g., using a triple-beam scale to weigh an object), to simple application of a well-practiced algorithm (e.g., adding two numbers), to complex procedures (e.g., implementing a multistep procedure to find out the density of an object). Procedural knowledge involves the knowledge of techniques and methods that are the result of consensus, agreement, or disciplinary norms (Anderson, et al., 2001). It involves how to do a task and is viewed as skill knowledge (Royer et al., 1993). Procedural knowledge is automatized over many trials (practice), allowing it to be retrieved and executed without deliberate attention. Automaticity is considered one of the key characteristics of procedural expertise (Anderson, 1983; Royer et al., 1993).

Schematic Knowledge (Knowing Why)

Schematic knowledge involves more organized bodies of knowledge, such as schemas, mental models, or "theories" (implicit or explicit) that are used to organize information in an interconnected and systematic manner. This form of organization allows individuals to apply principles or explanatory models to approach a problem; to provide an explanation (e.g., explaining why three same-sized blocks of different materials float, subsurface float, or sink in water); or to predict an outcome (e.g., whether an object will sink, float, or subsurface float according to its density and the medium's density). For example, schematic knowledge is involved, combined with procedural knowledge, in the process of reasoning from several theories to design experiments or to solve a new mathematical problem (De Kleer and Brown, 1983; Gentner and Stevens, 1983).

Strategic Knowledge (Knowing When, Where, and How to Apply Knowledge)

Strategic knowledge includes domain-specific strategies, such as ways to represent a problem or strategies to deal with certain types of tasks. It also entails general monitoring performance or planning strategies such as dividing a task into subtasks, reflecting on the process to explore alternative solutions, knowing where to use a particular piece of schematic knowledge, or integrating the three other types of knowledge in an efficient manner. It is important to mention that strategic knowledge—a higher-order knowledge—is based on the other three forms of knowledge (Anderson et al., 2001). An attempt to focus only on strategic knowledge without a strong base for the other forms of knowledge does not support transfer to new situations (Mayer, 1997; Pellegrino, 2002).

We can identify these types of knowledge in any content dimension (or subject matter domain). In science, mathematics, or technology education, we will always find facts, terminology, concepts, principles, methods, models, and strategies that together constitute the field. Differences in what curriculum developers, teachers, and assessors emphasize within a content domain will result in diverse curricula, instructional activities, and assessments, and, therefore, in different emphases on the types of knowledge.

What would be the importance of defining types of knowledge for assessing achievement? The types of knowledge framework proposed can help in the development (or selection) of assessments. The process is not straightforward. That is, there is not a perfect match between types of assessments and types of knowledge. However, when considering the cognitive demands involved in certain assessments, it seems possible to conclude that some types of knowledge may be better tapped by certain assessment tasks based on the affordances and constraints provided by those tasks (Li, Ruiz-Primo, and Shavelson, 2006).

Some Implications of the Achievement Framework for Assessment

The achievement framework has at least three implications. It helps to: (a) profile assessments (e.g., What is the achievement measured by a particular test?); (b) interpret students' scores (e.g., What exactly does a student's score represent?); and (c) design or select assessments (e.g., What types of assessment tasks can help us know whether students understand the concept of density?).

An example can help readers understand the first implication. In a study conducted by Li, Ruiz-Primo, and Shavelson (2006), items from the TIMSS 1999 Science Booklet 8 were analyzed. Results indicated that the TIMSS Booklet 8 science test is heavily loaded on declarative knowledge (approximately 50%), with the balance of procedural and schematic knowledge questions about equally distributed. Unfortunately, there were no example items for assessing students' strategic knowledge. The classification of the items was supported by a confirmatory factor analysis. Most factor loadings were statistically significant and generally high. The fit and the factor loadings support the feasibility of using our framework of science achievement to analyze the TIMSS science items. Analyzing large-scale or classroom assessments with an achievement framework in mind helps to identify not only the types of knowledge being assessed, but also to judge whether the assessments actually exemplify the standards measured (Shepard, 2003).

These results lead to the next implication. Think about students who have exactly the same total score in an assessment, but their response pattern is different. One student may arrive at the total score by answering correctly most of the items tapping declarative knowledge, whereas another one may respond correctly to those items tapping schematic knowledge. Therefore, using a single score to infer students' understanding can sometimes lead to untenable interpretations (Li, Ruiz-Primo, and Shavelson, 2006).

I explain the third implication in the next section on developing and evaluating assessments. At the Stanford Education Assessment Laboratory (SEAL), we have developed an approach to developing and evaluating assessments in science that uses the achievement framework, which we have named the "assessment square" (Shavelson, Ruiz-Primo, Li, and Ayala, 2002). This name distinguishes it from the "assessment triangle" (cognition, observation, and interpretation) proposed by Pellegrino et al., (2001).

TOWARD A FRAMEWORK FOR DEVELOPING AND EVALUATING ASSESSMENTS

In this section I briefly describe a modified version of the "assessment square" proposed by SEAL in 2002 (Shavelson, Ruiz-Primo, Li, and Ayala, 2002). The original assessment square has four main components (corners) presented in the following sequence: construct, assessment, observation, and interpretation. I have modified the assessment square by changing one of the components and the sequence of the elements (Figure 1): construct, observation models, assessment, and interpretation. A quote from Messick (1992) captures the rationale that guided my modification.

> [We] would begin by asking what complex of knowledge, skills, or other attributes should be assessed, presumably because they are tied to explicit or implicit objectives of instruction or are otherwise valued by society. Next, what behaviors or performances should reveal those constructs, and what tasks or situations should elicit those behaviors? Thus, the nature of the construct guides the selection or construction of relevant tasks as well as the rational development of construct-based scoring criteria and rubrics (p. 17).

The Assessment Square: A Quick View

The assessment square has four components, one at each corner of the four-sided figure (construct, observation model, assessment, and interpretation), and three types of analysis (conceptual, logical, and empirical). The arrows in the square intend to portray the idea that the development and evaluation of assessments involves an iterative process in which the corners of the square loop back to earlier corners.

Figure 1. The assessment square (adapted from Shavelson, Ruiz-Primo, Li, and Ayala, 2002).

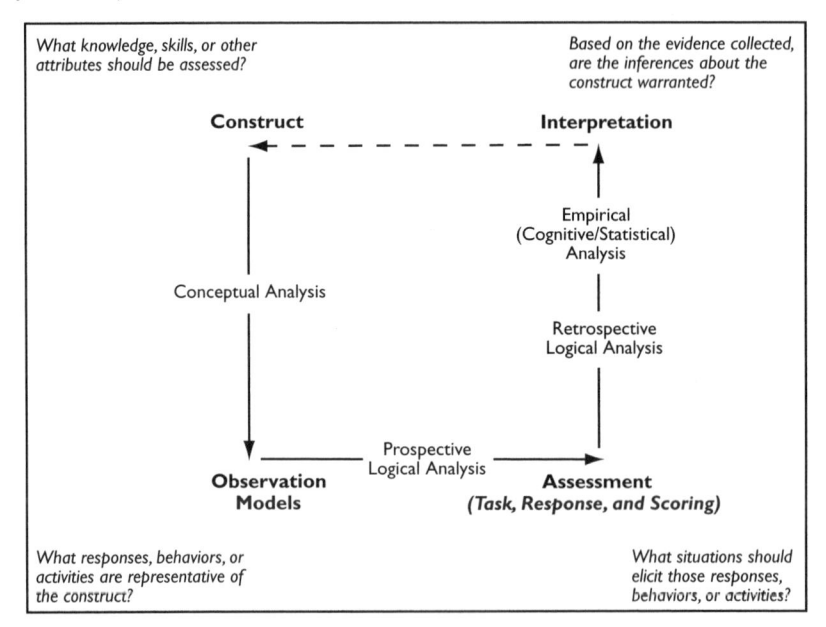

```
What knowledge, skills, or other          Based on the evidence collected,
attributes should be assessed?            are the inferences about the
                                          construct warranted?

        Construct  ←──────────────  Interpretation
            │                              ↑
            │                              │
            │                       Empirical
            │                       (Cognitive/Statistical)
            │                       Analysis
            │                              │
   Conceptual Analysis                     │
            │                       Retrospective
            │                       Logical Analysis
            │                              │
            │         Prospective          │
            ↓         Logical Analysis     │
        Observation  ──────────────→  Assessment
          Models                      (Task, Response, and Scoring)

What responses, behaviors, or             What situations should
activities are representative of          elicit those responses,
the construct?                            behaviors, or activities?
```

The assessment square begins with a working definition of what we are attempting to measure, the *construct*. At SEAL, we preferred the term construct, rather than cognition (Pellegrino, Chudowsky, and Glaser, 2001). A construct can be part of a theoretical model of a person's cognition (e.g., development of understanding of certain concepts in a domain), but does not limit assessment to an underlying cognitive model (Shavelson, Ruiz-Primo, Li, and Ayala, 2002; Shavelson and Ruiz-Primo, 1999).

The *conceptual analysis* focuses on the credibility of the link between the interpretive claim and the nature of the assessment (Ruiz-Primo, Shavelson, Li, and Schultz, 2001). The conceptual analysis helps describe the characteristics of the student knowledge upon which the inferences are to be based. It helps to delineate the tasks and response demands that the student will confront.

The conceptual analysis helps to define the *observation models*—the domain of tasks and responses that will be considered evidence of the construct. The observation models help to delineate what we will value in a student's response as evidence of the construct: In other words, what constitutes evidence about the construct in what is observed in any given assessment task (Mislevy et al., 1998).

The *logical analysis* focuses on the coherent link between the observation models and the nature of the assessment (Ruiz-Primo et al., 2001). Based on the conceptual analysis and the observation models, the nature of the assessment tasks and response demands that the student will confront can be logically determined. The logical analysis helps to describe the features of the tasks or situations that can be used to elicit the expected student performances. It is important to note that complex tasks often require individuals to apply more than one type of knowledge.

An *assessment* is a process of drawing reasonable inferences about what students know and are able to do on the basis of evidence derived from observing them in selected situations (Pellegrino et al., 2001). It involves a systematic procedure for eliciting, capturing or observing, and describing behavior, often with a numeric scale (Cronbach, 1990). An assessment can be thought of as a "sample" from a universe of possible assessments that are consistent with the construct definition (Shavelson, Baxter, and Gao, 1993; Shavelson and Ruiz-Primo, 1999). We have characterized assessments as having three components: a *task* (eliciting), a *response format* (capturing or observing), and a *scoring system* (describing behavior, possibly with a numeric value). We have argued that without these three components we do not have an assessment.

The third type of analysis, *empirical analysis,* involves collecting and summarizing students' behavior in response to the assessment task. Empirical analyses examine both the assessment-evoked cognitive activities and the students' observed performance. Empirical analyses, then, would include evidence of the following: (a) cognitive activities evoked as students perform the assessment, (b) the relationship between cognitive activities and performance levels (scores), and (c) statistical analyses necessary to determine the reliability and the different evidences necessary for a validity argument (Ruiz-Primo et al., 2001). It is expected that if we understand the link between assessments and cognitive processes, we should be able to design assessment tasks and responses to evoke different cognitive processes, different levels of performance, or both.

Finally, we put together evidence from the logical and empirical analyses and bring the evidence to bear on the validity of the *interpretations* from an assessment in relation to the construct it is intended to measure. We ask, "Are the inferences about the construct—a student's learning or domain knowledge—from performance scores warranted?" During the development of an assessment, we iterate through, somewhat informally, the assessment square until we have fine-tuned the assessment. In research and practice, where achievement and learning are assessed, we formally evaluate the inferences.

The Assessment Square: A Closer Look

What is the role of the achievement framework in the assessment square? The achievement framework helps us define the construct domain and the observation models, and to evaluate the quality of the assessment task. If the types of knowledge account for the different aspects of students' knowledge, then it is possible that constructs can be linked to types of knowledge. For example, a construct to be assessed in mathematics might be the fundamental concepts and procedures in algebra (Taylor and Bobbit Nolen, 2005). The construct is tapping declarative (knowledge of concepts in algebra) and procedural knowledge (knowledge of procedures in algebra).

Conceptual analysis then focuses on answering the question, "What does it mean to understand algebra concepts and procedures?" Understanding involves different cognitive processes, such as exemplifying, interpreting, classifying, summarizing, inferring, comparing, explaining, and implementing a procedure (Anderson et al., 2001). What in algebra is recognized as acceptable evidence of understanding a concept or a procedure? Conceptual analysis helps to define the observation models in more specific ways by focusing on how cognitive processes can be translated. For example, interpreting involves converting information from one representational form to another (paraphrasing words, converting numbers to words, and so on). Similar analyses can be done with other forms of understanding. This analysis helps to define the domain of tasks and responses (the observation models) that can be considered as evidence of the construct.

Would a constructed response (supply an answer) or a selected response (choose an answer) be more appropriate for assessing understanding of an algebra concept? The logical analysis helps us link the type of assessment to the type of knowledge by describing the features of the assessment tasks. Based on our research and the work of others (Li, 2001; Li, Ruiz-Primo, and Shavelson, 2006; Ruiz-Primo, Shavelson, Li, and Schultz, 2001), we have proposed four dimensions that can be used in logical analysis (Table 1): (a) *task demands*—what students are asked to perform; (b) *cognitive demands*—inferred cognitive processes that students likely act upon to provide responses; (c) *item openness*—degree of constraint in the nature and extent of the response; and (d) *complexity of the item*—diverse characteristics such as item familiarity, reading difficulty, and ancillary skills required.

Capabilities not explicitly part of what is to be measured, but nonetheless necessary for a successful performance, are called *ancillary* or *enabling* skills (Haertel and Linn, 1996). Ancillary skills have an impact on the characteristics of both the task and the format used by the student to respond. Both affect the level of difficulty of the assessment task. Therefore, ancillary skills have an impact on the validity of the assessment task. Observed variations in the scores of equally capable students due to differences in their ancillary skills give rise to construct-irrelevant variance in assessment scores.

Table 1. Assessment task dimensions.

Task Dimensions	Examples for Designing Tasks or for Coding Decisions
Task Demands: What the task asks the test taker to do; what it elicits from the student	• Define concepts • Identify facts	• Execute procedures in familiar tasks • Execute procedures in unfamiliar tasks	• Select an appropriate procedure to solve a problem • Determine the theoretical position of a manuscript • Draw diagrams illustrating a process • Write an equation from a statement • Construct an interpretation • Draw conclusions • Justify or Predict	• Evaluate the validity of a conclusion • Evaluate products, proposals • Produce alternative solutions to a given problem • Design an experiment to solve a non-routine or novel problem
Cognitive Demands: Inferred cognitive processes that students likely act upon to provide responses	Less Cognitive Demanding ←———————————————————→ More Cognitive Demanding Remembering / Recognizing / Recalling	Applying / Executing or Implementing more or less routine procedures	Reasoning Using Mental Models / Explaining / Interpreting / Inferring / Organizing/Classifying / Comparing/Contrasting / Exemplifying	Assembling Knowledge in New/Creative Ways / Planning / Generating / Producing / Monitoring
Item Openness Gradation in the constraint exerted in the nature and extent of the response	• The continuum from multiple-choice tests to constructed response test format such as long essays, projects, and collections of products over time • Require one correct solution versus multiple correct solutions/approaches • The continuum of the structuredness/directedness of the task (following instructions or steps)
Complexity Domain-general factors that influence the item difficulty	• Textbook-type task vs. ill-structured task • Inclusion of relevant and irrelevant information • Require only information found in task vs. information that can be learned from the task • Long reading demanding descriptions • Answers contradict everyday experience/belief • Cues provided or not

Task and cognitive demands are critical for establishing the link between assessment items and knowledge types, whereas item openness and complexity provide additional information to modify or revise the link (Li, Ruiz-Primo, and Shavelson, 2006). In other words, the item-knowledge link can be confirmed if all the information is consistent, or can be weakened if conflicting information is present. Logical analysis can be approached from two perspectives, for task development (prospective logical analysis) or for task selection/analysis (retrospective logical analysis). Prospective logical analysis should focus on the cognitive demands to help determine the characteristics of the assessment task. Retrospective analysis should focus first on the task demands.

Logical analysis leads to development or selection of an assessment (task, response format, and scoring system). For example, to tap understanding of algebra concepts and procedures, think about an assessment task that involves interpretation using the concept at hand as well as implementation of a procedure. The response format involves focusing on item openness and complexity. Making decisions about the format will depend on the context in which the assessment will be used (Figure 2). It is important to be aware that the decisions made about the response format will affect the complexity of the task. The characteristics of the scoring system will depend on the task demands and the characteristics of the response format. It is also important to consider that the type of item, by itself, does not necessarily reflect the complexity of the cognitive process involved.

Figure 2. Examples of items focusing on different types of knowledge: (a) declarative, (b) procedural, and (c) schematic.

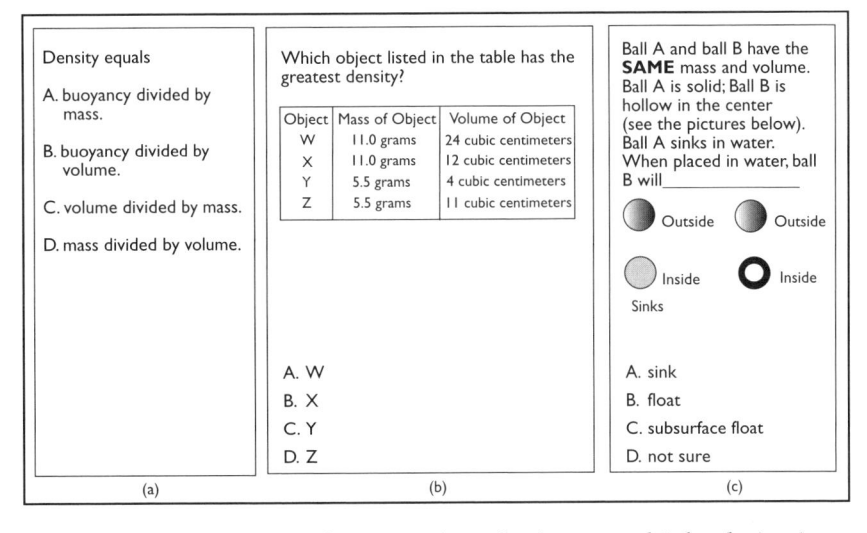

Density equals A. buoyancy divided by mass. B. buoyancy divided by volume. C. volume divided by mass. D. mass divided by volume.	Which object listed in the table has the greatest density? | Object | Mass of Object | Volume of Object | |---|---|---| | W | 11.0 grams | 24 cubic centimeters | | X | 11.0 grams | 12 cubic centimeters | | Y | 5.5 grams | 4 cubic centimeters | | Z | 5.5 grams | 11 cubic centimeters | A. W B. X C. Y D. Z	Ball A and ball B have the **SAME** mass and volume. Ball A is solid; Ball B is hollow in the center (see the pictures below). Ball A sinks in water. When placed in water, ball B will _____ Outside Outside Inside Inside Sinks A. sink B. float C. subsurface float D. not sure
(a)	(b)	(c)

Figure 2, which provides examples of science multiple-choice items used in a study on formative assessment (Shavelson and Young, 2000), illustrates the utility of retrospective logical analysis to classify assessment tasks into knowledge types. All three items focus on density. Example *a* can be thought of as an item tapping declarative knowledge. First, the response is expected to be in the form of a definition of a term (What is density?). This item asks a very specific content question, leaving students little opportunity to describe relationships between concepts or to apply principles. Second, the cognition evoked is likely to be retrieving information with a minimum of scientific reasoning. Third, in terms of item openness, the item is restricted in the sense that it is a multiple-choice item and the unfinished stem forces students to select options instead of responding to a question prior to reading any options. Being restricted, in turn, reinforces the task and cognitive demands placed on students. Finally, the coding for complexity did not add new information for the classification. Therefore, weighing the four characteristics, the item can be thought of as tapping declarative knowledge.

Example *b* was classified as tapping procedural knowledge. It provides students a table with information about the mass and volume of different objects. First, the item requires students to respond with an algorithm to calculate density and then to compare the different densities to arrive at the response. These two pieces of knowledge fall into the category of procedural knowledge. Second, the cognitive process students probably engage in is applying a calculation algorithm of dividing mass by volume to calculate density. Although this item allows students to generate their own responses before reading the options, students can arrive at the correct answer by working backward from the options, or even by merely guessing. The analysis of complexity does not provide additional information for modifying the coding decision.

Example *c* can be considered as tapping students' schematic knowledge. First, the item is asking for a prediction that should be based on an understanding of the concept of density. Second, the dominant cognitive process is reasoning with theories or mental models. It goes beyond the formula or how to apply it. An individual who can calculate density correctly every time may not respond to this item correctly if deeper understanding has not been reached. Finally, the item does not involve heavy reading or irrelevant information. The low complexity strengthens the posited link to schematic knowledge by reducing construct-irrelevant variances.

These examples make it clear that the logical analysis helps to analyze assessment tasks to define what types of knowledge may be tapped based on the affordances and constraints provided by those tasks. These examples also make it clear that the type of item does not necessarily reflect the complexity of the cognitive process involved. We acknowledge that linking assessment types to types of knowledge is not straightforward, since the testing method alone does not determine what type of knowledge an item measures (Bennett, 1993; Li, Ruiz-Primo, and Shavelson, 2006; Martinez, 1999; Snow, 1993). However, it is also true that there are some restrictions imposed by the nature of the type of assessment on its capacity to tap certain types of knowledge. For example, a typical multiple-choice item used to measure whether students understand "controlling variables" poses an experiment in the stem of the item and provides four or five experimental procedures, usually with drawings, that might be used to address it. The correct alternative is the only one that provides a clear comparison with respect to the critical variable. An item such as this can test whether students are able to detect relevant features in the options provided and to select the one that is considered the best instance. In a limited way, this type of item tests the students' understanding of controlling variables (Anderson et al., 2001), but it cannot probe the students' ability to analyze a situation and decide what variables need to be controlled (Haertel, 1991).

As mentioned, logical analysis does not allow for anticipating all the possible responses that the assessment will elicit from students (even competent students). The empirical analysis involves collecting and summarizing students' behavior in response to the assessment task. This analysis will provide evidence about the technical qualities of the assessment. We have proposed that, in addition to the more traditional strategies used to evaluate the quality of an assessment, an examination of the cognitive processes elicited by the assessment is also a critical source of information.

Cognitive analysis provides evidence about the cognitive activities that are evoked by the task as well as about the level of performance. We ask, "Does the assessment evoke the intended behaviors? Is there a correspondence between the intended behaviors and the performance scores?" Cognitive validity studies can provide information that is not provided by other psychometric methods. Cognitive validity can provide evidence of construct-relevant and -irrelevant sources of variance (Messick, 1995). For example, the characteristics of the assessment task may make students respond in ways that are not relevant (e.g., guessing), or may be so narrow that they fail to tap important aspects of the construct.

Several methods can be used to examine cognitive processes. Messick (1989) recommends using think-aloud protocols and retrospective reviews that ask examinees their reasons for providing incorrect answers. The usual procedure is to ask students to think aloud while completing an assessment item. Then, after completing an item, students respond to interview questions about the item. Talk-aloud protocols and interviews are audio taped and transcribed. In some studies (e.g., Kupermintz, Le, and Snow, 1999; Hamilton et. al., 1997), interviewers use a structured observation sheet to record events that cannot be captured on audiotape, such as the use of gestures. This information is added to the session transcripts. Another method, less intrusive, is to listen to students working in dyads talk to one another as they tackle an assessment task. These interactions provide real-time verbalizations of students' thinking and corresponding actions.

In evaluating achievement assessments we have thought of the assessment as a sample of student behavior (Shavelson and Ruiz-Primo, 2000; Shavelson, Baxter, and Gao, 1993). Inferences are made from this sample to a "universe" of behavior of interest. From this perspective, a score assigned to a student is but one possible sample from a large domain of possible scores that a student might have received if a different sample of assessment tasks were included, if a different set of judges were included, or if the assessment had occurred in a different context. Once a test score is conceived of as a sample of performance from a complex universe, statistical procedures can be brought to bear on the score's technical quality (including classical reliability theory, item response theory, and generalizability theory; Cronbach et al., 1972; Shavelson and Webb, 1991). However, a discussion of these statistical procedures is beyond the scope of this chapter.

Still, a perspective about assessment is not complete if the context in which these assessments are to be used is not considered. In the next section I will discuss issues around the context of assessment.

FUNCTION AND FORM IN ASSESSING STUDENTS

Any discussion about assessment needs to consider the purpose and the context, since both have implications for the design, implementation, and evaluation of the assessments (Haertel, 1991; Pellegrino et al., 2001; Shepard, 2003). Following Haertel's (1991) idea, let me pose different questions around assessment: Think of a student who asks her teacher why things sink or float, and the teacher responds. How does the teacher know whether the student actually understood his explanation? How can a teacher know whether his students understand the critical concepts just taught before moving on to the next instructional activity? At the end of a semester, how can a teacher determine whether students are moving forward in achieving the curriculum goals? How can a school district find out whether students are achieving the state standards? How can the state department of education monitor achievement trends over time? All these situations call for assessing students. However, the assessments differ in more than one way. For example, some assessments are *for* learning (formative), others *of* learning (summative, certification), and still others are used as accountability devices (Black, 1993; Black and Wiliam, 1998).

Assessment purposes and contexts matter for making decisions about the characteristics of the assessments. We can distinguish two broad assessment contexts: externally mandated and classroom-based assessments (Haertel, 1991). Different names have been used for these two contexts, but I use the terms "large-scale" and "classroom assessment," respectively, since they are the ones used more frequently in recent times (National Research Council, 2003; Pellegrino et al., 2001; Shepard, 2003). Large-scale assessments are designed to provide evidence about large numbers of students and may be implemented at the district, state, and national level. Classroom assessment, on the contrary, focuses on those assessments developed (or selected) and implemented by a teacher or group of teachers for use at the classroom or school level (Haertel, 1991). Focusing the discussion on the nature of assessments in these two contexts helps us understand key differences in purposes, and how those purposes affect the design, implementation, and evaluation of assessments.

Function and the Design of Assessments

No single assessment can fit all purposes. Furthermore, "the more purposes a single assessment aims to serve, the more each purpose will be compromised" (Pellegrino et al., 2001, p. 2). Three dimensions can be used to explain differences in the design of assessments for the two contexts described. These include: (a) remoteness of the assessments to classroom instruction, (b) extent of the content covered, and (c) constraints in the types of assessments used (Table 2).

Assessment in Science and Mathematics: Lessons Learned

Table 2. Comparing and contrasting the two contexts of assessment.

	DIMENSIONS	CONTEXTS	
		Large-Scale Assessment	Classroom Assessment
Design	Remoteness of the Assessment	Distal ←——————→	Immediate
	Extent of Content	Broad ←——————→	Narrow, tight, deep
	Constraints in Types of Assessment	More ←——————→	Less
Implementation	Frequency	Once a year ←——————→	Ongoing
	Requirements	Standardized ←——————→	Dynamic
	Closeness of Feedback	Delayed ←——————→	Immediate
Evaluation	Psychometric requirements	Stringent ←——————→	Flexible

Large-scale assessments are devised to measure a broad a range of content and, by design, cannot be tied to any one curriculum (Raizen et al., 1989). They "must necessarily be broad" in design (Shepard, 2003, p. 122) and, therefore, *distal* to classroom instruction (Ruiz-Primo et al., 2002). Think about defining the construct of what students know and can do at a given point in time. Usually, the content standards—district, state, national, or professional organization—will determine the knowledge and the skills to be assessed and, therefore, may not have a close relationship to what is happening in any given classroom. Often, there is "a gap between the objectives teachers and administrators would naturally develop, and those dictated by the inherently circumscribed nature of the external test" (National Research Council, 2003, p. 17). Furthermore, the sample of standards to be measured is somehow constrained by the assessment context. For example, a large-scale assessment task must be doable in the certain time provided; therefore, a standard focusing on communication skills in front of large groups is not very likely to be selected for a large-scale assessment.

Classroom assessments, on the other hand, must be as *close* as possible to what students are learning at any given time in order to guide the next instructional steps; that is, they should be based on the content and the activities of the enacted curriculum in any given classroom. Since classroom assessments are embedded in learning and teaching activities, the workable definition of the construct to be measured is narrower. In other words, it can focus on the concept taught in one particular day, in a lesson, or in a unit. Decisions on content to be sampled, then, are less striking. Since the constraints are less stringent, classroom assessments are less limited in the strategies that can be used to collect information about students' learning. For example, at an *immediate* level, classroom assessments are unobtrusive and informal. They can take the form of an informal observation or conversation, or they can be based on students' products. At the *close* and *proximal* levels, assessments can be more formal and planned. For formative purposes, embedded assessments can be implemented at critical points of the enacted curriculum to identify when students are not reaching the expected level or are holding misconceptions that impede learning. Embedded assessment can take almost any form (e.g., multiple-choice test, predict-observe-explain, concept maps, open-ended questions, students' products, or even planned classroom conversations). Assessments of complex performances are difficult to find in large-scale assessments, because such performances are easier to manage in the classroom setting (Taylor and Bobbitt Nolen, 2005).

Function and Implementing Assessments

There are at least three aspects of assessments that influence the implementation of assessments, including their requirements, their frequency, and the closeness of feedback following the assessment. For example, large-scale assessments are typically administered once a year. They usually take place at a predetermined, and limited, time. Standardization of procedures across testing sites is critical to ensure comparability across schools. Large-scale assessments for accountability or certification are often limited to evidence provided by a written test that is scored by agencies external to the classroom and the school (Black, 2003). Large-scale assessments can provide information directly, but not immediately, about individual students' achievement and can also provide information to teachers about a student's performance in particular areas (Shepard, 2003).

Classroom assessments, on the other hand, can happen on a continuous basis. They are local and contextual, and depend on the skills, knowledge, attentiveness, and priorities of teacher and students in any given classroom (Atkin and Coffey, 2003). They are more dynamic and do not require standardization in the same way large-scale assessment does (Shepard, 2003). Provision of feedback to the students is critical to the success of classroom assessment (National Research Council, 2003; Sadler, 1989; Shepard, 2003). When used for summative purposes, classroom assessments are more formal and focused (e.g., projects, term papers), and they can be close and/or proximal. At a close level, assessments focus within familiar contexts on the knowledge and skills students are being taught. At a proximal level, assessments focus on the same knowledge and skills, but they are embedded in contexts that are less familiar to the students. Proximal assessment tasks push students to extend the application of what they have learned—in other words, to transfer their knowledge.

Additionally, teachers are more in charge of classroom assessments. The evidence collected and the actions taken based on that evidence can have an impact at the next minute or in plans for the next year. Furthermore, some of the evidence may not require formal scoring but, rather, appropriate judgment. Since the main purpose of classroom assessment is improvement of learning, then assisted performance by the students is allowed (Shepard, 2003).

Function and Evaluation of Technical Quality of Assessments

Purpose and context also determine the assessment's technical requirements. Three key factors have been mentioned as critical for the technical quality of assessments: validity, reliability, and fairness. Standards for technical accuracy are quite different for large-scale than for classroom assessments, but both must meet their respective standards in these areas (National Research Council, 2003). Due to the potential consequences attached to large-scale assessment results, which may be used for accountability, certification, and internal school decisions about tracking and grading, the technical quality of these assessments assume greater importance than those used in classroom assessments (Black, 2003; Shepard, 2003). The evidence collected and procedures used to provide evidence about the validity and reliability of large-scale assessments tend to be more complex and require more evidence than evaluations of the technical quality of classroom assessments. Also, the clinical judgment of teachers plays a more important role in classroom assessment (Shavelson et al., 2003), which tends to focus on formative assessment issues: maintaining a clear view of the learning goals, considering information about the state of students' performance, and taking actions to close performance gaps (Sadler, 1989).

FINAL COMMENTS ABOUT THE ACHIEVEMENT FRAMEWORK AND THE ASSESSMENT SQUARE

The examples presented in this chapter focus on assessment tasks that could exemplify items for embedded classroom or for large-scale assessment (for formative or summative purposes). I focus now on *informal formative assessment*—assessment that generates evidence of learning during the course of daily classroom activities and that can take place at any level of student-teacher interaction, whether whole class, small group, or one-on-one (Bell and Cowie, 2001; Duschl, 2003; Shavelson, Black, Wiliam, and Coffey, 2003). We (Ruiz-Primo and Furtak, 2004, in press) have adopted the term *assessment conversation* (Duschl and Gitomer, 1997; Duschl, 2003) to refer to these daily instructional dialogues that embed assessment into an activity already occurring in the classroom. Assessment conversations permit teachers to recognize students' conceptions, mental models, strategies, language use, or communication skills in order to guide instruction.

Assessment conversations center on questions that help to make students' thinking explicit. Assessment conversations can also tap different types of knowledge. The quality of questions teachers ask the students verbally and informally during a classroom discussion, or more formally in written formats to guide the students' discussions, can also be designed with different cognitive demands and tapping different types of knowledge. For example, a teacher may ask "What is density?" (declarative knowledge focusing on recalling information), or he can ask, "How do you know that?" or "What evidence do you need to support your explanation of density?" (schematic knowledge focusing on explanations). Questions in assessment conversations that tap knowledge at higher levels can allow teachers to gather better information about students' levels of understanding than questions that just focus on recalling information—or worse, on simple answers with low/null cognitive demands (e.g., Teacher: "Is density a property of the objects?" Student: "Yes").

Like formal assessment tasks, informal assessments can derive from a construct (What knowledge and skills do I want my students to learn?), from observation models (Which aspects of student performance are the most important to focus on? How will students demonstrate their knowledge and skills?), or from assessments (What questions or tasks will help me to know whether or not my students can solve mathematical problems?). In informal classroom assessment, scoring can be informal. It is probably obvious that reliability for informal assessment is not possible to calculate. However, it is important that teachers always think about whether the informal prompts they use reflect the learning goals, whether they draw out the targeted knowledge and skills, and whether they can be used to compare students' responses across different types of questions and tasks. Consistency of students' responses across tasks provides evidence of the validity of the inferences made based on the information collected.

CONCLUSION

In this chapter, I have described a way to conceptualize achievement and assessment. Achievement has been conceptualized as *types* of knowledge (declarative, procedural, schematic, and strategic) that represent different levels of organization of knowledge. I have also described the assessment square as a strategy for developing, selecting, and evaluating assessments. The strategy has four components: construct to be measured, observation models, assessment, and inferences. I have also discussed two contexts of assessments, large-scale and classroom assessments, which determine how assessments are designed, implemented, and evaluated.

At SEAL, we have built assessments for both large-scale and classroom-embedded science assessments, which have been developed or analyzed based on the conception of achievement presented (Stanford Education Assessment Laboratory [SEAL], 2003). This conceptualization has provided a lens not only for building assessments, but also for analyzing the curriculum or standards on which a large-scale assessment is based. We have used the assessment framework for guiding the evaluation of the developed assessments. The framework has led to a more accurate profile of large-scale assessments in terms of types of knowledge tapped, and to the design of assessments that better tap diverse types of knowledge. Furthermore, it has been made apparent that some curricula lead students to the development of declarative and procedural knowledge, but not to schematic and strategic knowledge. The implementation of the framework has provided enough evidence for curriculum developers to agree with our conclusions (SEAL, 2003). The assessment framework has led to the design of a logical analysis strategy for analyzing assessment tasks, and has enabled us to focus on issues that have proved to be important sources of information in assessing validity (Ruiz-Primo et al., 2001). There is a hope that the strategies proposed in this chapter will prove to be as useful to members of the technology education community as they have been to us (SEAL) in analyzing the quality of assessments.

DISCUSSION QUESTIONS

1. What do we mean by "achievement" in technology education?
2. What are the issues that we need to consider in developing or selecting assessments to measure achievement in technology education?
3. What types of assessments should we use to assess whether a person is technologically literate?

REFERENCES

Alexander, P. A., and J. E. Judy (1988). The interaction of domain-specific and strategic knowledge in academic performance. *Review of Educational Research* 58(4): 375–404.

American Educational Research Association, American Psychological Association, and National Council on Measurement in Education (1999). *Standards for educational and psychological testing.* Washington, DC: American Educational Research Association.

Anderson, J. R. (1983). *The architecture of cognition.* Cambridge, MA: Harvard University.

——— (1997). *Cognitive psychology and its implications*, 4th ed. New York: W. H. Freeman and Company.
Anderson, L. W., D. R. Krathwohl, P. W. Airasian, K. A. Cruikshank, R. E. Mayer, P. R. Pintrich, J. Raths, and M. C. Wittrock (2001). *A taxonomy for learning, teaching, and assessing: A revision of Bloom's taxonomy of educational objectives.* New York: Longman.
Atkin, J. M., and J. E. Coffey (2003). *Everyday assessment in the science classroom.* Arlington, VA: National Science Teachers Association.
Baxter, G. P., A. D. Elder, and R. Glaser (1996). Knowledge-base cognition and performance assessment in the science classroom. *Educational Psychologist* 31(2): 133–140.
Bell, B., and B. Cowie (2001). *Formative assessment and science education.* Dordrecht, Netherlands: Kluwer Academic Publishers.
Bennett, R. E. (1993). On the meaning of constructed response. In *Construct versus choice in cognitive measurement*, edited by R. E. Bennett and W. Ward, 1–27. Hillsdale, NJ: Lawrence Erlbaum Associates.
Black, P. (1993). Formative and summative assessment by teachers. *Studies in Science Education* 21:49–97.
——— (2003). The importance of everyday assessment. In *Everyday assessment in the science classroom*, edited by J. M. Atkin and J. E. Coffey, 1–11. Arlington, VA: National Science Teachers Association.
Black, P., and D. Wiliam (1998). *Assessment and classroom learning.* Assessment in Education 5(1): 7–74.
Bybee, R. W. (1996). The contemporary reform of science education. In *Issues in science education*, edited by J. Rhoton and P. Bowers, 1–14. Arlington, VA: National Science Teachers Association.
Chi, M. T. H., P. Feltovich, and R. Glaser (1981). Categorization and representation of physics problems by experts and novices. *Cognitive Science* 5(2): 121–152.
Chi, M. T. H., R. Glaser, and M. Farr, eds. (1988). *The nature of expertise.* Hillsdale, NJ: Lawrence Erlbaum Associates.
Cronbach, L. J. (1990). *Essentials of psychological testing*, 5th ed. New York: Harper Collins.
Cronbach, L. J., G. C., Gleser, H. Nanda, and N. Rajaratnam (1972). *The dependability of behavioral measurements.* New York: Wiley.

de Jong, T., Fergusson-Hessler (1996). Types and qualities of knowledge. *Educational Psychologist* 31(2): 105–113.

De Kleer, J., and J. S. Brown (1983). Assumptions and ambiguities in mechanistic mental models. In *Mental models,* edited by D. Gentner and A. L. Stevens, 155–190. Hillsdale, NJ: Lawrence Erlbaum Associates.

Duschl, R. A. (2003). Assessment of inquiry. In *Everyday assessment in the science classroom,* edited by J. M. Atkin and J. E. Coffey, 41–59. Washington, DC: National Science Teachers Association Press.

Duschl, R. A., and D. H. Gitomer (1997). Strategies and challenges to changing the focus of assessment and instruction in science classrooms. *Educational Assessment* 4(1): 37–73.

Gelman, R., and J. G. Greeno (1989). On the nature of competence: Principles for understanding in a domain. In *Knowing, learning, and instruction: Essays in honor of Robert Glaser,* edited by L. B. Resnick, 125–186. Hillsdale, NY: Lawrence Erlbaum Associates.

Gentner, D., and A. L. Stevens, eds. (1983). *Mental models.* Hillsdale, NJ: Lawrence Erlbaum Associates.

Glaser, R., and E. Silver (1994). Assessment, testing, and instruction: Retrospect and prospect. In *Review of research in education* 20, edited by L. Darling-Hammond, 393–419. Washington, DC: American Educational Research Association.

Haertel, E. (1991). Form and function in assessing science education. In *Science assessment in the service of reform,* edited by G. Kulm and S. M. Malcom, 233–245. Washington, DC: American Association for the Advancement of Science.

Haertel, E., and R. L. Linn (1996). Comparability. In *Technical issues in large-scale performance assessment,* edited by G. W. Phillips, 59–78. Washington DC: National Center for Education Statistics.

Hamilton, L. S., E. M. Nussbaum, and R. E. Snow (1997). Interview procedures for validating science assessments. *Applied Measurement in Education* 10(2): 181–200.

Kupermintz, H., V. H. Le, and R. Snow (1999). Construct validity of mathematics achievement: Evidence from interview procedures. *CSE Technical Report 493.* Los Angeles, CA: UCLA/National Center for Research on Evaluation, Standards, and Student Testing (CRESST).

Li, M. (2001). *A framework for science achievement and its link to test items.* PhD diss., Stanford University, Stanford, CA.

Li, M., M. A. Ruiz-Primo, and R. J. Shavelson (2006). Towards a science achievement framework: The case of TIMSS 1999. In *Contexts of learning mathematics and science,* edited by S. J. Howie and T. Plomp, 291–311. London, England: Routledge.

Li, M., and R. J. Shavelson (April 2001). Using TIMSS items to examine the links between science achievement and assessment methods. Paper presented at the annual meeting of the American Educational Research Association, Seattle, Washington.

Martinez, M. E. (1999). Cognition and the question of test item format. *Educational Psychology* 34(4): 207–218.

Mayer, R. (1997). Incorporating problem-solving into secondary school curricula. In *Handbook of academic learning: Construction of knowledge,* edited by G. D. Phye, 473–492. San Diego, CA: Academic Press.

Messick, S. (1984). Abilities and knowledge in educational achievement testing: The assessment of dynamic cognitive structures. In *Social and technical issues in testing: Implications for test construction and usage,* edited by B. S. Plake, 155–172. Hillsdale, NJ: Lawrence Erlbaum Associates.

——— (1989). *Validity. In Educational Measurement,* edited by R. Linn, 201–219. New York: Macmillan Publishing Company.

——— (1992). The interplay of evidence and consequences in the validation of performance assessments. *Educational Researcher* 23(2): 13–23.

——— (1995). Validity of psychological assessment: Validation of inferences from persons' responses and performances as scientific inquiry into score meaning. *American Psychologist* 50(9): 741–749.

Mislevy, R. J., L. S. Steinberg, F. J. Breyer, R. G. Almond, and L. Johnson (1998). A cognitive task analysis, with implications for designing a simulation-based performance assessment. *CSE Technical Report 487.* Los Angeles, CA: UCLA/National Center for Research on Evaluation, Standards, and Student testing (CRESST).

National Research Council (2003). *Assessment in support of instruction and learning: Bridging the gap between large-scale and classroom assessment.* Washington, DC: National Academies Press.

Organisation for Economic Co-operation and Development (2003). *The PISA 2003 assessment framework: Mathematics, science and problem-solving knowledge and skills.* OECD Programme for International Student Assessment.

Pellegrino, J. W. (February 2002). Understanding how students learn and inferring what they know: Implications for the design of curriculum, instruction, and assessment. Paper presented at the National Science Foundation Instructional Materials Development Principal Investigators' Conference, Reston, VA.

Pellegrino, J. W., N. Chudowsky, and R. Glaser (2001). *Knowing what students know: The science and design of educational assessment.* Washington, DC: National Academies Press.

Raizen, S. A., J. B. Baron, A. B. Champagne, E. Haertel, I. V. S. Mullis, and J. Oakes (1989). *Assessment in elementary school science education.* Andover, MA: The Network Inc.

Royer, J. M., C. A. Ciscero, and M. S. Carlo (1993). Techniques and procedures for assessing cognitive skills. *Review of Educational Research* 63:201–243.

Ruiz-Primo, M. A. (1997). *Toward a framework of subject-matter achievement assessment.* Stanford University, Stanford, CA.

——— (February 2002). On a seamless assessment system. Paper presented at the Seamless Science Education Symposium, AAAS Annual Meeting, Boston, MA.

Ruiz-Primo, M. A., and E. M. Furtak (2004). Informal assessment of students' understanding of scientific inquiry. Paper presented at the American Educational Research Association Annual Conference, San Diego, CA.

——— (In press). Exploring teachers' informal formative assessment practices and students' understanding in the context of scientific inquiry. *Journal of Research in Science Teaching.*

Ruiz-Primo, M. A., Shavelson, L. Hamilton, and S. Klein (2002). On the evaluation of systemic science education reform: Searching for instructional sensitivity. *Journal of Research in Science Teaching* 39(5): 369–393.

Ruiz-Primo, M. A., R. J. Shavelson, M. Li, and S. E. Schultz (2001). On the validity of cognitive interpretations of scores from alternative concept-mapping techniques. *Educational Assessment* 7(2): 99–141.

Sadler, D. R. (1989). Formative assessment and the design of instructional systems. *Instructional Science* 18(2): 119–144.

Shavelson, R. J., G. P. Baxter, and X. Gao (1993). Sampling variability of performance assessments. *Journal of Educational Measurement* 30(3): 215–232.

Shavelson, R. J., P. Black, D. Wiliam, and J. E. Coffey (2003). On aligning formative and summative functions in the design of large-scale assessment systems. Paper presented at the National Research Council's Assessment in Support of Instruction and Learning: Bridging the Gap between Large-Scale and Classroom Assessment Workshop, Washington, DC.

Shavelson, R. J., and M. A. Ruiz-Primo (1999). On the assessment of science achievement (English version), *Unterrichts wissenschaft* 27(2): 102–127.

——— (2000). On the psychometrics of assessing science understanding. In *Assessing science understanding: A human constructivist view*, edited by J. J. Mintzes, J. H. Wandersee, and J. Novak, 303–341. Orlando, FL: Academic Press.

Shavelson, R. J., M. A. Ruiz-Primo, M. Li, and C. C. Ayala (2002). Evaluating new approaches to assessing learning. CSE Technical Report 604. Los Angeles, CA: UCLA/National Center for Research on Evaluation, Standards, and Student Testing (CRESST).

Shavelson, R. J., and N. M. Webb (1991). *Generalizability theory: A primer.* Newbury Park, CA: Sage.

Shavelson, R. J., and D. Young (2000). *Embedding assessments in the FAST curriculum: On beginning the romance among curriculum, teaching and assessment.* Stanford University, Stanford, CA.

Shepard, L. A. (2003). Reconsidering large-scale assessment to heighten its relevance to learning. In *Everyday assessment in the science classroom*, edited J. M. Atkin and J. E. Coffey, 121–146. Arlington, VA: National Science Teachers Association.

Snow, R. E. (1993). Construct validity and constructed-response tests. In *Construction versus choice in cognitive measurement: Issues in constructed response performance testing and portfolio assessment*, edited by. R. E. Bennett and W. C. Ward, 45–60. Hillsdale, NJ: Lawrence Erlbaum Associates.

Stanford Education Assessment Laboratory (2003). On the integration of formative assessment in teaching and learning with implications for teacher education. Paper presented at the EARLI annual meeting, Padova, Italy.

Sugrue, B. (1995). A theory-based framework for assessing domain-specific problem-solving ability. *Educational Measurement: Issues and Practices* 14(3): 29–36.

Taylor, C. S., and S. Bobbitt Nolen (2005). *Classroom assessment: Supporting teaching and learning in real classrooms.* Upper Saddle River, NJ: Pearson-Merrill Prentice Hall.

Third International Mathematics and Science Study (TIMMS) (2003). *TIMSS assessment framework and specification 2003.* International Association for the Evaluation of Educational Achievement (IEA).

White, R. (1999). *The nature and structure of knowledge: Implications for assessment.* (n.p.).

Yin, Y. (2005). The influence of formative assessments on student motivation, achievement, and conceptual change. PhD diss., Stanford University, Stanford, CA.

Skills Assessment in Trade and Industrial Education

Chapter 12

Janet Z. Burns
Karen M. Schaefer
Georgia State University

INTRODUCTION

To apply skills assessment in trade and industrial (T&I) education, it is necessary to consider not only the nuts and bolts of assessment itself, but also the part T&I education plays in preparing society's future workforce. T&I education takes its raw material—the beginning student—and from that raw material molds a finished product, a skilled worker or an advanced student. One way the success or failure of this educational process is measured is through assessment techniques. However, to look at assessment before viewing its place in the systems that maintain it and the activities that precede it puts the cart before the horse. Assessment in T&I education can only be sensibly addressed by considering what pulls it. Long before educators can begin to design appropriate assessments for their T&I programs, they must identify the purpose and desired performance outcomes that comprise the driving force behind their assessment procedures. Identifying the components of a T&I program can only be done meaningfully, in conjunction with the needs of business and industry.

The goals of this chapter are (a) to explain how a T&I program's purpose, objectives, and assessments are interrelated; (b) to examine what is assessed in T&I programs; (c) to consider the design of various assessment instruments used in T&I programs; and (d) to relate assessment to T&I program accountability.

HOW ARE A T&I PROGRAM'S PURPOSE, OBJECTIVES, AND ASSESSMENTS INTERRELATED?

To maintain relevance, T&I education must reflect the continually developing needs of work organizations. As industry needs and requirements change and cultural assumptions alter, job-related criteria will adjust in turn. For example, in recent decades, as computer-based occupations have spread throughout the workplace, technical skill qualifications for employees have become more complex. A static skill set may have sufficed in the past, but today's work environment requires versatile problem-solving abilities. In today's work world, employees with flexibility, adaptability, and ingenuity may be prized more than those with fixed abilities, regardless of how masterful they are (DTI Associates Inc., n.d.). Additionally, as the workforce culture has diversified, employers seek assurance that would-be employees possess a productive work ethic (Hill, 2005). Employers will continue to require employees with adequate technical and skill competencies to meet new and evolving industry standards but will also seek employees whose social competencies and positive work attitudes enable them to perform effectively in diverse and changing work environments. It is the responsibility of T&I education to ensure that the competencies taught in T&I programs match the needs of today's business and industry.

To accomplish this goal, T&I programs must implement a systematic approach in designing and evaluating the total process of learning and teaching. This approach must link with industry to determine and integrate the T&I program's purpose, its specific performance objectives, and its selection of appropriate assessment methods—all of which combine to bring about valid and effective instruction. Like tongue-and-groove joints, these various components of the T&I program are both integral and interrelated. Missing one, the T&I program will not have its basic building materials, and if the components do not fit snugly together, the program will fail to cohere. According to Morrison, Ross, and Kemp (2004), questions that need to be answered before one can create appropriate assessments include the following:

- For whom is the program developed (purpose)?
- What do you want the students to learn or demonstrate (objectives)?
- How is the content best learned (teaching methods)?
- How do you determine the extent to which the learning is achieved (assessment)?

These four fundamental components—purpose, objectives, methods, and assessment—form the framework for systematic instructional planning (Morrison, Ross, and Kemp, 2004).

T&I Program Purpose

Although different T&I programs may have different foci, the program exists to fill certain needs determined by society, business and industry, and/or local school systems. Typically, there are three predominant purposes that T&I education serves. Many T&I programs in the secondary school setting combine two or even all three of these purposes.

At its most elementary level, T&I education has the purpose of career exploration. At this level, students in the program are provided with the opportunity to investigate a variety of T&I career areas in broad terms. This allows the students to examine their aptitude and interest in a career field before they invest the time and the cost of further training in a postsecondary institution, whether it is a technical school, college, or university.

A second purpose of T&I education is to prepare workers for entry-level employment. When this is the focus, students enrolled in T&I programs often plan to go directly into the workforce after graduating from their secondary schools. For this reason, T&I educators with this as their program's aim must align their training programs to closely articulate with industry needs through methods such as the implementation of advisory boards, mentorships, job-shadowing programs, and other close associations between students, T&I teachers, and industry specialists.

At its most advanced level, a T&I program's purpose is to form a partnership between a secondary and postsecondary institution to promote a seamless education in which students transition smoothly from the secondary school to the postsecondary institution for advanced training. This is accomplished through arrangements such as Tech Prep, in which students receive credit at the postsecondary institution for courses studied in the secondary school classroom and, after graduating from high school, enroll in the postsecondary institution for further training (U.S. Department of Education, 2006). Integral to this process is cooperation and communication among all the participants, namely, students, teachers at both the secondary and postsecondary institutions, and experts from business and industry. Currently, 7,400 high schools, or approximately 47% of the nation's high schools, offer at least one Tech Prep program. Nearly every technical and community college in the United States participates in a Tech Prep consortium. Other participants include four-year colleges and universities, employer and union organizations, and private businesses (U.S. Department of Education, 2006).

T&I Program Objectives

Performance objectives state what a learner is expected to know, do, and value after instruction. Only after a T&I program's purpose is clarified is it possible to define the program's performance objectives. It is the purpose of the T&I program that dictates the expectations of what knowledge students in the program will learn, what skills they will master, and what attitudes they will adopt.

With the final component of the learning process—assessment—in mind, the performance objectives are expressed using clearly measurable terms so that meaningful assessment becomes possible. For example, a performance objective in a T&I program whose purpose is career exploration might be to "identify 10 different careers in the construction industry." An objective in a program whose purpose is to prepare students for entry-level employment might be to "interpret the symbols on a blueprint." If the purpose of the program is preparation for postsecondary education, a performance objective might be to "create a blueprint using computer-aided design software." In each case, the statement of the objective, using concrete verbs such as "identify," "interpret," or "create," makes it possible to design a valid assessment that determines whether a student mastered the objective.

Just as the performance objectives hinge upon a clearly defined purpose, so too do learning activities evolve from clearly stated performance objectives. Only after the performance objectives are determined can meaningful instruction be designed. Instructional methods, classroom exercises, and student projects are all focused on achieving the results outlined by the performance objectives.

T&I Program Assessment Methods

After the purpose is defined, the performance objectives are determined, and the instruction has occurred, it is assessment that reveals whether or not the T&I program has accomplished its goals. Assessment might be considered the final phase of T&I education, and, like the Olympic race after all the years of preparation and training, assessment is the phase in which the T&I program proves its worth. It is assessment that provides assurance to the businesses and industries that make use of the program's final "product" that a T&I program is successfully preparing employees capable of fulfilling their business needs.

The assessment measures the extent to which the performance objectives have been met. If the objectives are designed to comply with the T&I program's purpose, are planned with the needs of business and industry in mind, and are written in terms of outcomes that can be meaningfully measured, they can be reliably examined. In this final phase, the four components of the learning process—purpose, objectives, teaching methods, and assessment—come together in a unified picture of the T&I program.

WHAT IS ASSESSED IN T&I PROGRAMS?

Intellectual and Problem-Solving Skills

T&I programs must encompass a broad range of instruction that calls for the teaching of knowledge, training of skills, and molding of workplace attitudes. These three areas of learning are often referred to as the categories (or domains) of cognitive, psychomotor, and affective learning. Each of these domains is further organized within a hierarchy of learning levels from simple to complex.

In designing valid assessment instruments, a T&I teacher must take care to address each of these three domains of learning, as well as ensure that the assessments encompass the spectrum of learning levels. It is essential that teachers prepare objectives and assessments at all levels of the hierarchies, sequencing the teaching and testing of objectives from simple to more complex. Objectives at lower levels of complexity must be mastered in order for students to achieve more complex behaviors (Borich, 2004). Alexander (1996) points out that when task-relevant knowledge or skills necessary for acquiring more complex behaviors have not been taught, students may demonstrate high error rates and less active engagement in the learning process at the higher levels of behavioral complexity. Additionally, categorizing learning into separate cognitive, affective, and psychomotor domains does not mean that learning in one domain is not needed for learning in other domains (Anderson et al., 2001). For example, participating in a laboratory activity requires not only thought (cognitive domain), but operating a tool or piece of equipment (psychomotor domain), and having an attitude to safely complete the activity (affective domain). Although a single learning objective often refers to behaviors from only one of the three domains, one or more behaviors from the other domains often are necessary for the dominant learning behavior to occur.

Within the cognitive domain, learning may occur from the lowest level, in which a learner memorizes terms or facts, to the highest level, in which a student solves complex and novel problems. To assess learning at the lowest level, a student might, for example, be required to label parts of an engine. At the top of the hierarchy, the teacher assesses problem-solving abilities. Here the teacher might present the student with a variety of malfunctioning engine symptoms to analyze and require him or her to diagnose the problem and repair the defective part.

As emerging technologies and employer expectations have evolved, the intellectual skills of the cognitive domain have become increasingly emphasized in T&I programs. During the industrial age of the early 1900s, education created a system that best met the workplace needs of industrial manufacturing and assembly lines. Schools educated students with skills to match those jobs. As the needs of society have changed to make use of new technologies, the role of education has changed so that students are now being prepared to live, learn, and work in an international community (Edwards, 1998). Formerly, achieving the facility and dexterity of a master craftsman may have placed a large proportion of the goals of vocational training in the psychomotor domain, but to be successful in today's workforce additional skills in problem solving are needed in such areas as entrepreneurship, marketing, teamwork, leadership, sales, and writing business plans (Georgia Department of Education, 2006).

Process and Product

Despite the movement within T&I training toward more emphasis on cognitive skills, in many T&I programs the largest percentage of learning still occurs in the psychomotor domain. Assessment in this domain is centered on evaluation of processes and products. Although closely related, processes and products are not the same. A process is a method of doing something and often involves a number of ordered steps. Its aim is to complete a specified task or to produce a particular product. Its assessment determines to what extent students demonstrate mastery of the steps required to produce a product, but does not appraise the product itself. A product, on the other hand, is the completed outcome, or result, of a process. To assess a product, the teacher evaluates the degree of completion and the quality of the finished piece.

The distinctions between the assessments of processes and products are (a) the object of the evaluation and (b) the time frames in which the scoring or evaluation occurs. When the focus is on the process—the "doing"—scoring examines the method of performance and occurs at the same time as the performance. For example, in a culinary arts class, a teacher might evaluate students as they mix batter for a cake to determine if they are following the appropriate steps in the correct order. Steps that might be evaluated are such things as whether the student used the correct measuring spoons, added the ingredients in the proper order, or mixed the batter to the right consistency. In contrast, when evaluating cake baking as a product rather than a process, the quality of the baked and completed cake is assessed. At this time, the teacher might evaluate qualities such as whether the cake has evenly risen, completely baked, and has a light, fluffy texture.

Attitudes and Personal Qualities

The affective domain relates to personal qualities, values, attitudes, and, of particular importance to T&I fields, interpersonal relationships. Hill states that "workplace supervisors and team leaders value workers who possess a positive work ethic" (Hill, 2005, p. 5), yet "the increasingly diverse workforce and the challenges this diversity brings to the traditional camaraderie of work" lead to potential areas of conflict and differences in work ethic standards (p. 17). Today, more than ever, it is the task of a T&I training program to design assessment measures that will assure prospective employers that the T&I education graduates have acquired work-ethic traits that enable them to be successful employees. Some of the items to assess in the affective domain are a student's capacity to work cooperatively in team settings, ability to resolve conflicts, willingness to complete tasks or assignments, and tendency to attend class regularly and on time.

HOW DOES A T&I TEACHER DESIGN ASSESSMENT INSTRUMENTS?

Criteria for Selecting an Assessment Method

Assessment is the culminating element of the four fundamental components that frame instructional planning. To select appropriate assessment methods for a T&I program, it is once again necessary to consider its three prior components, that is, the T&I program's purpose, its objectives, and its teaching methods.

The selection of an assessment method is, in large part, directed by the purpose and objectives of the T&I program. The intent of the objective and the choice of action verbs used in the statement of the objective often dictate the appropriate type of assessment. An objective that requires a student to "identify" objects or "define" terms may prescribe a matching or short answer paper-and-pencil-type assessment, whereas an objective that asks a student to "create" a product calls for a more authentic or hands-on assessment technique. The difficulty level of the assessment must correspond to the purpose of the program and therefore depends on whether its purpose is to provide the students with an opportunity for career exploration, to prepare the students for entry-level employment, or to enable the students to transition smoothly from the secondary school to an institution of advanced training.

Ideally, an assessment should not only supply a source of scores for grading the students but also provide a teaching tool that enhances the students' learning opportunities. This is more likely to be accomplished by using rubrics to assess processes or projects, rather than by using traditional paper-and-pencil tests (Burns, 2005; Wiggins, 1989). When a teacher shares rubrics with students prior to instruction and assessment and provides students with the opportunity to discuss and use the rubrics for self-checks, students are empowered to take charge of their own learning. The students become more focused and self-directed, find learning and assessment less threatening, and are more reflective about their learning (Custer, 1996). Because students vary in their preferred learning styles, the teacher should also make use of a variety of assessment methods. In addition to accommodating students' differing learning styles, the use of an array of assessment instruments allows for evaluation at various levels in the learning hierarchy.

While many of the criteria for selecting assessment methods are related to the fundamental components of instructional planning, others are strictly pragmatic and concern issues of cost, time, and feasibility. Expense may be a consideration if the assessment procedure requires providing the students with materials, which is often the case with skills demonstrations and other authentic assessments. Time may also be a limiting factor, not only the time involved in administering the assessment, but also the time involved in preparing the assessment instrument. Lack of appropriate or sufficient laboratory equipment may also prohibit the use of some assessment techniques.

Aligning the Assessment Instrument to the Learning Domain

T&I teachers must evaluate knowledge learned, skills mastered, and attitudes adopted. Each of these areas of learning tends to lend itself to specific assessment methods. Therefore, once the teacher has selected the objectives to be assessed and determined the assessment's appropriate difficulty level, he or she must also choose an assessment form that aligns with the domain of learning that is to be evaluated.

The cognitive domain, which involves factual knowledge, intellectual reasoning, and problem solving, has traditionally been assessed with conventional paper-and-pencil testing techniques. At the lower levels of the cognitive hierarchy, paper-and-pencil methods such as matching items, simple multiple choice, true/false questions, and completion items are often used, while short-answer, more complex multiple-choice, restricted essay, and unrestricted essay methods can be used to assess skills at a higher cognitive level. Recently, educators have advocated a move away from paper-and-pencil tests when assessing higher-level cognitive skills, instead replacing them with authentic assessments. (Readers may refer to chapters 3 and 4 for more detailed discussions of traditional and performance assessments.) These include such methods as analyzing case studies, synthesizing ideas by writing papers, and reflecting and evaluating through the creation of portfolios (Ward and Murray-Ward, 1999; Taggart and Wood, 1998). While authentic forms of assessment can furnish a more comprehensive picture of students' abilities, devising a reliable scoring system for them may present a challenge. This difficulty can often be surmounted by the use of scoring rubrics. Four particularly useful rubric formats for T&I that are discussed in this chapter are checklists, rating scales, score cards, and employability charts.

Checklists, rating scales, and score cards also provide a means of assessing processes, products, and other skills which fall largely in the psychomotor learning domain. A checklist is a list of characteristics, criteria, or steps that the teacher expects the student to exhibit, apply, or follow. It requires a simple "check" of whether or not each item on the list has been addressed. That is, for each item on the list, the student has either fulfilled that requirement or the student has not (see Figure 1). A teacher can easily customize a checklist to suit the skill he or she wishes to assess. The list of characteristics might describe the required features of a finished product, or the steps might outline the necessary actions to be taken in a process.

Figure 1. Example of a checklist for a process (using a microscope).

Microscope Checklist (Student Actions)	Yes	No
1. Takes slide.		
2. Places drop of culture on slide.		
3. Places slide on stage.		
4. Looks though eyepiece keeping other eye closed.		
5. Turns to low-power objective.		
6. Turns to high-power objective.		
7. Keeps eye to eyepiece while adjusting.		
8. Removes slide from stage.		
9. Wipes objective with lens paper.		
10. Wipes eyepiece with lens paper.		
Total Score		

A checklist can also serve as a teaching method in and of itself. Students, provided in advance with a checklist, can follow its steps as they learn to complete a process, or they can use the checklist to evaluate and reflect on their own or each other's completed projects.

A checklist is simple to create and straightforward to apply, but because quality is not appraised, it provides only a quick, surface evaluation. For a more in-depth evaluation, a rating scale may be more appropriate.

Like a checklist, a rating scale lists a set of characteristics. However, unlike a checklist, accompanying each characteristic on the rating scale is a set of multilevel descriptors, with a specific point value assigned to each level, based on the quality of work exhibited in each category (see Figure 2). The rating-scale format of a rubric is ideal for assessing student attitudes and behaviors. Because of the multilevel descriptors, students can readily see the lowest or most undesirable type of attitude or behavior, as well as the top level. This aids students in knowing what is expected of them. It is possible

Skills Assessment in Trade and Industrial Education

to accomplish both self-evaluation and cooperative evaluation using this format (Burns, 2005). Although a rating scale is more difficult to devise than a checklist, this addition of a scoring system gives a rating scale an advantage over a simple yes-no checklist. Rating scales assess not just whether a characteristic is present or whether a procedure has been followed, they also allow for a nuanced numeric evaluation of how fully the characteristic is exhibited or how accurately the procedure is performed.

Figure 2. Excerpt from a rating scale: "How I Rate as a Team Player."

1	2	3	4	5	Rating
COOPERATION					
I never follow directions. I do not work well with others. I am never on task or focused.		I follow directions sometimes. I work well with others sometimes. I am on task most of the time. I sometimes remain focused.		I always follow directions. I work will with others. I always stay on task. I always remain focused.	
CONTRIBUTIONS TO TEAM					
I do not contribute. I am unwilling to work.		I contribute only when directed or contribute infrequently. I am sometimes willing to work.		I contribute frequently, needing little direction. I am willing to do the work.	
RELIABILITY					
I am never on time. I never have material. I am never accountable. I never finish work.		I am usually on time. I usually have material. I am usually accountable. I usually finish work.		I am always on time. I always have material. I am always accountable. I always finish work.	
COMMUNICATION SKILLS					
I am narrow-minded. I do not share ideas. I monopolize the conversation. I ignore team members.		I am usually open-minded. I sometimes share ideas and let others share. I sometimes listen.		I am open-minded. I share ideas and let others do the same. I always listen to others.	

A scorecard is a hybrid of a checklist and a rating scale. Like a checklist, a scorecard lists a set of characteristics, and the teacher places a check mark to indicate whether each characteristic has or has not been demonstrated by the student. At the same time, however, the scorecard provides a fixed point value for each item on the list, thus making it possible to assign a student a numeric score (see Figure 3). Scorecards can be integrated into the learning process by allowing students to help develop the score card for a specific assignment, including the list of characteristics to be placed on the score card and the point values to be assigned to each characteristic. When scorecards are used as a means of self- or cooperative evaluation, they function both as an assessment system and as an additional teaching method.

Figure 3. Example of a scorecard: Scorecard for a business-writing product.

Scoring Rubric	Standard Score	My Score	Teachers Score
1. Ideas Well-defined Clear presentation Supporting arguments	15		
2. Organization Logical presentation of topics Conclusion follows from details	10		
3. Development All details relevant Variety in sentence structure	10		
4. Conventions Appropriate language for intended audience Grammatical constructions Spelling	20		
5. Document Format	15		
Total	70		

Traits that fall in the affective domain of learning can be particularly troublesome to score with impartiality. Nevertheless, T&I teachers, as they prepare their students for the workforce, may need to assess their students' attitudes and personal qualities, particularly in areas that apply to work ethics. By adapting checklists and scorecards for this purpose, subjectivity can be minimized and fairness and reliability increased. One example of this adaptation is an "employability chart," which incorporates a variation of a scorecard and makes it possible to assign numeric scores to personal traits that concern a would-be employer. Like the rating scale, the employability chart describes levels of each behavior listed. Rather than allotting points for characteristics demonstrated by a student, an employability chart may work in reverse: all students begin with 100 points and points are deducted for behaviors that detract from employability. By subtracting points for not following directions, inefficient use of time, cheating, excessive absences, lack of group participation, nonadherence to dress codes, use of profanity, or other unprofessional conduct, the resulting "employability score" can assess values, such as honesty and integrity; work-ethic qualities, such as dependability and initiative; and interpersonal skills, such as cooperativeness and courtesy.

In cases where the teacher wants a general understanding of a student's viewpoints and attributes rather than a numeric score, attitudes and personal qualities can be assessed informally through the use of questionnaires, student journaling, and personal conferences.

HOW ARE T&I ASSESSMENT STRATEGIES AND PROGRAM ACCOUNTABILITY RELATED?

Student assessment, which takes place shortly after learning has occurred, determines the degree to which the program's learning objectives have been met by each individual student. Program accountability, on the other hand, determines the effectiveness of the entire program—its curriculum, its resources, its procedures, its activities, and the overall competence of its student graduates. To some degree, program accountability can only be determined after graduates have moved on to business and industry or to postsecondary educational settings.

Like all school programs, T&I programs are accountable to their local school systems, their state departments of education, and, under the No Child Left Behind legislation, the federal government. In many cases, accountability to these agencies is ensured through standardized testing or through on-site reviews by outside accreditation agencies, such as the various regional branches of the Association of Colleges and Schools.

T&I programs, however, in contrast to other secondary school programs, have additional accountability requirements. Because T&I programs receive funding through the Carl D. Perkins Act, they must adhere to the Perkins accountability guidelines. Perhaps most importantly, to merit continuance, T&I programs must assure business and industry that the skills they teach are relevant and that the students they graduate are able to perform as capable and ethical professionals. Some T&I programs provide this assurance in part by preparing their students for industry licensure. Because licensure requirements are set by industry, independent of the secondary school program, when a T&I program's students successfully meet the license requirements, the program receives independent corroboration of its efficacy. Other T&I programs demonstrate their accountability by fulfilling the requirements and standards set by business and industry for industry certification. These industry-certified T&I programs receive credentials, once again from sources outside the school system, that validate and confirm their program's relevance.

All T&I programs, like other educational programs, must collect data on an ongoing basis to assess program effectiveness. Regardless of how a T&I program is assessed, a viable T&I program must be flexible and able to revise its purpose, goals, and objectives as business and industry requirements and conditions change in today's dynamic economic environment.

CONCLUSION

A T&I program's purpose, objectives, instructional methods, and assessments are interrelated and must be coordinated for greatest effectiveness. Its purpose determines its objectives, which, in turn, guide both instruction and assessment procedures. Education in T&I fields involves the teaching of knowledge, development of skills, and molding of attitudes. To evaluate learning in each of these three domains requires the use of a variety of assessment instruments. Although traditional paper-and-pencil testing may serve to assess the simple memorization of facts, evaluating higher-level competencies, such as the completion of a process or the creation of a product, may require the use of authentic assessment techniques. Used in conjunction with authentic assessments, checklists, rating scales, scorecards, and employability charts aid in eliminating scoring subjectivity. T&I programs can demonstrate their relevance and validity through student licensure, industry certification, or program assessment.

Knowledge of assessment is equally important to teachers in all fields. T&I and technology education teachers must be well versed in evaluating learning in the cognitive, psychomotor, and affective domains. Both program areas are aimed at developing the "whole individual" with hands-on experiences and career exploration opportunities. In addition to knowledge testing, teachers in both areas must assess laboratory experiences, processes and products, and attitudes or work ethic. While the curriculum focus of technology education is different than T&I, teachers in both program areas need to employ a variety of assessment techniques to gain a more complete picture of student learning and of program effectiveness.

DISCUSSION QUESTIONS

1. How do your assessments match your program's purpose?
2. How do you think learning objectives dictate your selection of assessment techniques for your program?
3. How could you as a teacher use a checklist, rating scale, scorecard, and employability chart in your program?
4. What do you think is the value of assessing your students in all three domains of learning?
5. What do you think are some advantages and disadvantages of authentic assessment techniques versus paper-and-pencil testing?
6. How do national standards, state standards, and industry or workforce development standards affect how you as a teacher design assessments in your classroom?

REFERENCES

Alexander, P. (1996). The role of knowledge in learning and instruction. *Educational Psychologist* 31:89–145.
Anderson, L. W., D. R. Krathwohl, P. W. Airasian, K. A. Cruikshank, R. E. Mayer, P. R. Pintrich, J. Raths, and M. C. Wittrock (2001). *Taxonomy for learning, teaching, and assessing: A revision of Bloom's taxonomy of educational objectives.* New York: Longman.
Borich, G. D. (2004). *Effective teaching methods.* 5th ed. Upper Saddle River, NJ: Pearson Education Inc.
Burns, J. Z. (2005). *Evaluation strategies for teachers and trainers,* n.p.
Custer, R. L. (1996). Rubrics: An authentic assessment tool for technology education. *The Technology Teacher* 55:27–37.

DTI Associates Inc. (n.d.). *Charting a new course for career and technical education.* http://www.ed.gov/about/offices/list/ovae/pi/hsinit/papers/cte.pdf (accessed September 18, 2006).

Edwards, E. (1998). Assessment that drives instruction. In *Rubrics: A handbook for construction and use,* edited by G. L. Taggard, S. J. Phifer, J. A. Nixon, and M. Wood, 22. Lanham, MD: The Scarecrow Press Inc.

Georgia Department of Education (2006). *Reengineering CTAE for the 21st century.* Atlanta, GA: Georgia Department of Education.

Hill, R. B. (2005). Work ethic characteristics: Perceived work ethics of supervisors and workers. *Journal of Industrial Teacher Education* 42(2): 5–20.

Morrison, G. R., S. M. Ross, and J. E. Kemp (2004). *Designing effective instruction.* 4th ed. New York: John Wiley & Sons Inc.

Smith, P. L., and R. J. Tillman (2005). *Instructional design.* 3rd ed. Hoboken, NJ: John Wiley & Sons Inc.

Stevenson, C. (2002). *Teaching ten to fourteen year olds.* 3rd ed. Boston: Allyn and Bacon.

Taggart. G. L., and M. Wood (1998). Rubrics: A cross-curricular approach to assessment. In *Rubrics: A handbook for construction and use,* edited by G. L. Taggard, S. J. Phifer, J. A. Nixon, and M. Wood, 5774. Lanham, MD: The Scarecrow Press Inc.

U.S. Department of Education (2006). *Tech prep education and high school improvement.* http://www.ed.gov/about/offices/list/ovae/pi/cte/tpreptopic2.html (accessed September 15, 2006).

Waks, L. J. (2002). Exploratory education in a society of knowledge and risk. In *Middle school curriculum, instruction and assessment,* edited by V. A. Anfara Jr., and S. L. Stacki, 23–40. Westerville, OH: National Middle School Association and Information Age Publishing.

Ward, A. W., and M. Murray-Ward (1999). *Assessment in the classroom.* Belmont, CA: Wadsworth Publishing Company.

Wiggins, G. (1989). A true test: Toward more authentic and equitable assessment. *Phi Delta Kappan* 70:703–713.

Conducting Program Assessments

Chapter 13

Gerald F. Day
University of Maryland Eastern Shore

Anthony E. Schwaller
St. Cloud State University

CHAPTER GOALS AND OBJECTIVES

This chapter focuses on assessment of technology teacher education programs. The overall purpose of this chapter is to present acknowledged principles, practices, and content used to assess total technology education programs, a process known as program assessment. More specifically, this chapter is designed to
- analyze the 10 major principles of program assessment;
- compare different program assessment styles, measures, and techniques;
- identify various program assessment organizations;
- determine principles, procedures, and content used to assess the instructional design of technology education programs;
- determine principles, procedures, and content used to assess teacher candidates and faculty in technology education programs;
- identify additional program assessment components and strategies for teacher candidates and faculty in technology education programs; and
- identify the principles and procedures used to assess physical facility components of technology education programs.

Although many of the principles and strategies described here could be applied to program assessment in any type of educational setting, the primary emphasis will be on assessment of technology teacher education programs at colleges or universities.

PRINCIPLES OF PROGRAM ASSESSMENT IN TECHNOLOGY EDUCATION

Program assessment is defined as the process of assessing educational effectiveness and student achievement in regards to the total technology education program. To assess any technology education program adequately, it is necessary to base the assessment on a set of sound and logical assessment principles (Banta et al., 1996). Understanding the following 10 assessment principles will help technology educators better assess their technology education programs. Each principle is presented and defined. In addition, several questions are presented to help technology education faculty better discuss, define, and analyze each of these principles during any program assessment process.

Assessment Principle One: The assessment of student learning begins with educational values. Program assessment is not an end in itself. It is a means to improve instruction, student achievement, and the overall technology education program. Its effective practice begins with and endorses a vision of the kinds of learning we most value for technology education students. Educational values should drive not only "what" we choose to assess in technology education, but also "how" we assess it.

Questions to be discussed among the faculty to address this principle of program assessment should include:
- Why should program assessment be considered a means to an end?
- How do faculty members define the process of learning?
- Why is the "how" of a learning experience so important?
- Which is more important, the "what" or the "how" of program assessment in technology education?

Assessment Principle Two: Assessment is most effective when it reflects an understanding of learning as multidimensional, integrated, and revealed in student performance over time. Learning in technology education is a complex process. It entails not only what technology education students know, but also what they can do with this knowledge. It also involves student values, attitudes, dispositions, and habits of mind. Standards-based curriculum design and assessment efforts attempt to define what the disciplinary knowledge, abilities, and dispositions should be. All of these components affect both academic success as well as performance beyond the classroom.

Program assessment should reflect this assessment principle by employing a diverse array of assessment methods, including those that call for actual performance. These should be used over time so as to measure change, growth, and increasing degrees of integration of technological subject matter.

Questions to be discussed among the faculty to address this principle of program assessment should include:
- What types of assessments should be used for our program assessment?
- Why should a variety of assessments be used?
- Why should program assessment be continued over time?

Assessment Principle Three: Assessment works best when the program it seeks to improve has clear, explicitly stated purposes. Program assessment in technology education should be goal-oriented. It entails comparing educational performance with educational purposes and expectations. These should be derived from the institution's mission, vision, and instructional framework; from the department's mission, vision and goals; and from overall course objectives and design.

Questions to be discussed among the faculty to address this principle of program assessment should include:
- Why should program assessment be tied to the goals of the program?
- What is the relationship between the institution's mission statement, the department's mission and goals, and specific course objectives?
- What is the relationship between the stated mission, goals, and objectives and the assessment practices we are using?

Assessment Principle Four: Assessment requires attention not only to outcomes but also and equally to the experiences that led to those outcomes. Information about student learning outcomes in technology education is of very high importance. In other words, it matters greatly where technology students "end up." However, to improve the technology education learning experience, we need to know about student experiences along the way. We need to know about the curricula, teaching styles, and kinds of student efforts that led to particular technology education outcomes and standards. In most technology education program assessments, the outcomes are directly related to the *Standards for Technological Literacy: Content for the Study of Technology* (International Technology Education Association [ITEA], 2000).

Questions to be discussed among the faculty to address this principle of program assessment should include:
- Why is it important to know how students get to where they are going?
- How do the technology education curriculum, teaching methods, and student efforts and experiences lead to a particular outcome?

Assessment Principle Five: Assessment works best when it is ongoing and continuous, not episodic. Program assessment is a cumulative process. Although isolated or one-time event assessment can be better than none, program assessment is best fostered when it entails a linked series of assessment activities completed over time. This means tracking the progress of students over time and collecting data about program activities and outcomes on an ongoing basis.

Questions to be discussed among the faculty to address this principle of program assessment should include:
- Why should program assessment be continuous and cumulative?
- What happens if program assessment is isolated or a "one-time event"?
- How will a continuous and cumulative approach to program assessment affect the way our technology education program operates?

Assessment Principle Six: Assessment fosters wider improvement when representatives from across the educational and business communities are involved. Student learning is a school-wide and community responsibility. Program assessment strategies should recognize and reflect that shared responsibility. Thus, while program assessment efforts may start small, the goal over time is to involve a variety of educators, businesspeople, and other members of the community. Program assessment should certainly include others from outside of the field of technology education.

Questions to be discussed among the faculty to address this principle of program assessment should include:
- Why should program assessment be both a campus and a community-wide responsibility?
- What are the advantages of having broader involvement from the community?
- What other groups or individuals should, or could, be involved with our program assessment process?

Assessment Principle Seven: Assessment makes a difference when it begins with critical issues and illuminates questions that people really care about. Program assessment in technology education recognizes the value of information to help in the process of improvement. But to be useful, program assessment practices should be connected to issues or questions that technology education teachers really care about. This can often include standards developed by the larger professional community, such as the *Standards for Technological Literacy.*

Questions to be discussed among the faculty to address this principle of program assessment should include:
- Why is it important to connect program assessment to issues that teachers really care about?
- What are examples of issues that concern technology education teachers?

Assessment Principle Eight: Assessment is most likely to lead to improvement when it is part of a larger set of conditions that promote change. Just conducting a program assessment process in technology education may change little. The greatest contribution of program assessment within any technology education program comes when it is part of a wider program improvement initiative, such as national accreditation or campus reform.

Questions to be discussed among the faculty to address this principle of program assessment should include:
- What should be the relationship between program assessment and teaching and learning in technology education?
- How does classroom assessment relate to the quality of teaching and learning in technology education?
- What are the national goals, objectives, or standards the institution is trying to attain?

Assessment Principle Nine: Through assessment, educators meet responsibilities to students and to the public. More than ever, there is today a compelling and important public stake in any education system. As technology education faculty we have a responsibility to the public that supports and depends on us to provide pertinent technological knowledge and skills to the students for their success in the future.

Questions to be discussed among the faculty to address this principle of program assessment should include:
- Should technology educators be concerned about public opinion of their programs?
- Do we have a responsibility to the public to teach to the *Standards for Technological Literacy?*
- What relationship exists between accountability and program assessment?
- How does a strong program assessment process in technology education contribute to improved accountability?

Assessment Principle Ten: Assessment is most effective when undertaken in an environment that is receptive, supportive, and enabling. Successful technology education program assessment requires an environment characterized by effective leadership, administrative commitment, adequate resources (clerical support, funding, etc.), faculty and staff development opportunities, and time to carry out the program assessment process. Furthermore, to achieve the desired program improvement goals that typically drive program assessment, these assessment activities must be approached openly and without fear of castigation, which might otherwise cause negative findings to be hidden or whitewashed.

Questions to be discussed among the faculty to address this principle of program assessment should include:
- Does the leadership within our institution support program assessment and program improvement activities?
- What other institutional resources are available to assist in the program assessment process?
- Is there sufficient faculty and staff development to effectively implement program assessment?

PROGRAM ASSESSMENT MEASURES AND TECHNIQUES

Both formative (usually focuses on program improvement) and summative (usually focuses on evaluation for accountability) assessments are typically incorporated into the overall program assessment process. The following examples of assessment techniques represent some of the more common methods used to perform program assessment.

State Licensure Tests

Most states require some type of standardized licensure test for state teaching certification, such as PRAXIS (basic test of mathematics, reading, and writing) and PRAXIS II (technology education specialty test of pedagogy and content). Student scores on these tests are one measure used to determine the quality of the program and how effective it is in preparing teacher candidates. Such tests are considered summative assessments and are often part of the program assessment process.

Placement Rates

This type of summative assessment technique is often used to determine if the graduates of the program are being placed in careers related to the objectives and goals of the program.

Exit Interviews

This type of assessment technique is used to gather information and to show overall student achievement from teacher candidates who have just completed a learning experience such as a course, a sequence of courses, or a complete technology education program. It can serve either a formative or a summative purpose.

Student GPA

This type of assessment is considered a summative assessment technique used to determine the teacher candidate's overall level of competence after completion of the technology education program. Program GPA assessment can be used for program assessment and reflects the overall level of teacher candidate achievement and skills in technology education.

Student Portfolios

This type of assessment is a collection of artifacts such as projects, papers, activities, assignments, lesson plans, curriculum units, educational philosophy, résumés, and awards that students have developed and completed throughout their entire technology education program. Most of today's portfolios are electronic rather than paper-based. Electronic portfolios allow the use of live-stream video to provide a visual record of teacher candidate performance. This type of assessment technique could be formative or summative, depending upon when the portfolio is assessed and how it is used in the program.

Evaluation by Practicum Advisor

This type of assessment technique is usually completed during a specific learning experience such as an internship, clinical experience, or student teaching in order to alter or improve the experience, or to show teacher candidate achievement in technology education. This type of assessment is generally considered a formative style of assessment.

Focus Groups

This type of assessment involves a meeting between students and faculty in small groups with a set of questions designed to determine the level of competence, teacher candidate achievement, and candidate growth as the students progress through the program. This type of assessment technique is typically considered formative and can certainly be used to enhance the overall program assessment of technology education programs.

Graduate Exit Survey

Generally considered a summative assessment process, graduate surveys are used to identify graduate perceptions of the program and the overall quality and relevance of the technology education learning experience. Graduate surveys are usually conducted one to five years after graduation.

Capstone Experiences

This type of program assessment usually occurs during a senior seminar or final course or experience in the technology education program. This type of program assessment is designed to integrate all of the learning experiences that have taken place within the program. It is typically a summative form of assessment.

Employer Survey

Generally, a survey is sent to employers who hire graduates, inquiring about the employee's (graduate's) quality, competence, capabilities, dispositions, and skill levels. Like graduate surveys, the trend for employer surveys is to send and complete them through e-mail using an online form. Usually this type of assessment measures the employer's satisfaction with the worker's skills and dispositions. This type of assessment is considered summative.

Other Assessments

Other assessment techniques are also common in technology education program assessments. Some of the more popular techniques include oral examinations, pretests and posttests, standardized tests, peer assessments, rubrics and checklists, work samples, videos, student interviews, and computerized assessments (ITEA, 2003). These and other strategies are useful for gathering data that can assist in evaluating program quality, structure, delivery strategies, faculty, and other program features that may be included in the program assessment process.

PROGRAM ASSESSMENT ORGANIZATIONS

The National Council for Accreditation of Teacher Education (NCATE)

The National Council for Accreditation of Teacher Education (NCATE), the largest national teacher education accreditation organization, is a national accrediting body for schools, colleges, and departments of education whose goals are to establish and maintain high-quality teacher preparation. Through a meticulous assessment process, NCATE determines which schools, colleges, and departments of education meet rigorous national standards in preparing teachers and other school personnel such as principals and school counselors. Through this process of professional accreditation, NCATE strives to make a difference in the quality of teaching and teacher preparation (National Council on Accreditation of Teacher Education [NCATE], 2006).

Five groups were instrumental in creating NCATE in 1954. These include the American Association of Colleges for Teacher Education (AACTE), the National Association of State Directors of Teacher Education and Certification (NASDTEC), the National Education Association (NEA), the Council of Chief State School Officers (CCSSO), and the National School Boards Association (NSBA). They recognized the need for a strong, independent, quality assurance mechanism composed of all key stakeholders in education.

NCATE's accreditation process is based on six standards:
- Standard 1: Candidate knowledge, skills, and dispositions
- Standard 2: Assessment system and unit evaluation
- Standard 3: Field experiences and clinical practices
- Standard 4: Diversity
- Standard 5: Faculty qualifications, performance, and development
- Standard 6: Unit governance and resources

The educational or professional unit on a campus as a whole is accredited through an assessment process using these six standards. If the institution has a technology education program, the ITEA and Council on Technology Teacher Education (CTTE) Specialized Professional Association (SPA) evaluates and awards national recognition to that program separately. The ITEA/CTTE SPA is the recognized professional association for technology teacher education programs. The ITEA/CTTE SPA has its own set of 10 standards for technology education, which have been approved by NCATE. These 10 standards, like the standards for other SPAs, help the department or professional unit on a campus to meet NCATE's Standard 1 (candidate knowledge, skills, and dispositions). Technology education departments submit a program report electronically that contains assessment data addressing the 10 standards. More information can be obtained on the CTTE and NCATE Web sites (NCATE, 2006).

Regional Accreditation Organizations

Each region in the United States has a regional accreditation organization. These organizations assess all teacher education programs in the educational unit as a whole. Examples of these regional accreditation organizations would be the Middle States Commission on Higher Education (MSCHE) and the Southern Regional Education Board (SREB). The SREB, the nation's first multi-state compact for education, works to improve every aspect of education—from early childhood to doctoral degrees and beyond. The SREB compiles information, shares resources, evaluates teacher education programs, develops demonstration programs, and conducts conferences. Goals include assurance that every student is taught by qualified teachers, and that activities related to teacher quality crosscut many of its activities (Southern Regional Education Board, 2006). The MSCHE evaluates degree-granting institutions of higher education based on standards developed by the colleges and universities that are members of the Middle States Association of Colleges and Schools. It accredits institutions that meet its standards and assists institutions in improving their programs and services (Middle States Commission on Higher Education, 2006). Other regional accreditation organizations provide similar services to the SREB and the MSCHE.

The National Board for Professional Teaching Standards (NBPTS)

The National Board for Professional Teaching Standards' (NBPTS) mission is to advance the quality of teaching and learning by maintaining high and rigorous standards for what highly qualified teachers should know and be able to do, by providing a national voluntary system certifying teachers who meet these standards, and by advocating related education reforms. The certification process involves the development of an electronic teaching portfolio based on standards. To date, the NBPTS has developed standards in 27 fields and a process for assessing accomplishments for each. One of the fields is Career and Technical Education, which includes as a cluster the area of technology education. National Board certification complements, but does not replace, state licensing, and depending on state and school district policies, offers individuals opportunities for more pay and greater mobility across states. The NBPTS also collects data related to Board certification and supports a research program that investigates the impact of National Board certification on teachers, students, public policies, and educational reform (National Board for Professional Teaching Standards, 2006).

State Standards

In addition to national and regional organizations involved in program assessment and accreditation, many state departments of education have established state standards that are used to evaluate different educational programs. These state standards are often used in conjunction with national standards to assess a program's effectiveness. For example, although some technology teacher education programs elect not to go through the entire formal NCATE review and on-site visitation process, their state department of education may have established a partnership with the NCATE. In such states, known as "Partnership States," specific state standards have been reviewed by the SPAs to determine if their state standards align with the national SPA standards. Thus, passing the state's accreditation review assures that the program meets SPA guidelines (NCATE, 2006).

ASSESSING INSTRUCTIONAL CONTENT IN TECHNOLOGY EDUCATION PROGRAMS

Technology education is a discipline designed to promote technological literacy. These programs should be designed to produce individuals who can solve problems involving the technical means humans use for their survival. Technological literacy is considered a basic and fundamental goal for all students, regardless of their educational or career goals.

Assessing the instructional content in technology education should be based on the most up-to-date national and state content standards. This section will focus on five of the 10 national ITEA/CTTE SPA standards that were approved by NCATE in 2004. These five content standards were developed by the CTTE Accreditation Committee through a rigorous process of development and review. Each standard includes a variety of knowledge, performance, and disposition indicators to help in the assessment process.

The document *Standards for Technological Literacy: Content for the Study of Technology* (ITEA, 2000) includes an area entitled "The Designed World" which focuses on technologies found in medical, agricultural, biotechnical, energy, communication, transportation, construction, and manufacturing contexts. It is not necessary to have courses in each of these areas in a technology education program, but evidence of teacher candidate performance in these areas should be collected to indicate the comprehensiveness of the program's curriculum.

The first subject matter standard, "The Nature of Technology," states that technology education program teacher candidates should be able to develop an understanding of the nature of technology within the context of the designed world. The technology education program should prepare candidates who can explain the characteristics and scope of technology and compare the relationship among technologies and the connections between technology and other disciplines. They should be able to apply the concepts and principles of technology in a classroom and laboratory setting. Candidates should comprehend the nature of technology in a way that demonstrates sensitivity to the positive and negative aspects of technology in our world.

The second standard, "Technology and Society," states that the technology education candidate should be able to develop an understanding of the interactions between technology and society. This includes such knowledge as the ability to compare the relationships between technology and social, cultural, political, and economic systems. This standard also suggests that the technology education candidate should have the ability to assess the importance of significant technological innovations on the history of humankind. Candidates should be able to judge the effects of technology on the environment and evaluate the relationship between technology and social institutions such as the family, religion, education, government, and the workforce.

The third standard, "Design," indicates that the teacher candidate should develop an understanding of design within the technological contexts that are part of the designed world. Candidates should be able to explain the importance of design in the human-made world, describe the attributes of technological design, and analyze engineering design processes and principles. They should be able to apply the processes of troubleshooting, research and development, invention, innovation, and experimentation in developing solutions to a design problem. Candidates should be able to investigate the relationship between designing a product and the impact of the product on the environment, economy, and society.

The fourth standard, "Abilities for a Technological World," states that the teacher candidate should be able to develop abilities within the context of the designed world. This standard covers a variety of knowledge, performance, and disposition attributes. Among these is that candidates should demonstrate an ability to operate and maintain technological products and systems and analyze designed products by identifying the key components of how they work and how they are made. Candidates should be able to evaluate design solutions and develop suggestions for design improvements. They should also be able to develop, model, and refine design solutions. Moreover, candidates should be able to operate technological devices and systems, including diagnosing and restoring a malfunctioning system. Safe practices and procedures in the use of tools and equipment should be displayed throughout this standard.

The fifth standard, "The Designed World," involves the ability to analyze the principles, concepts, and applications of technology within all of the designed world contexts. This includes selecting appropriate technologies and effectively using them in a variety of contexts.

These five standards form one basis for evaluating the instructional content in teacher education programs. State standards may also have to be integrated if the respective state department of education has established additional instructional or content standards. The emphasis within evaluation of instructional content is focused on teacher candidates' demonstrated knowledge and their capacity to advance student learning in a clinical or student teaching experience.

ASSESSING PEDAGOGICAL SKILLS OF TEACHER CANDIDATES IN TECHNOLOGY EDUCATION PROGRAMS

The remaining ITEA/CTTE SPA standards focus on pedagogical skills. The sixth standard, "Curriculum," deals with the teacher candidate being able to design, implement, and evaluate curricula based upon the *Standards for Technological Literacy: Content for the Study of Technology* (ITEA, 2000). This involves identifying appropriate content for the study of technology at different grade levels; integrating curriculum content from other fields of study, such as mathematics and science; and identifying curriculum and instructional materials that enable effective instructional delivery. The teacher candidate should be able to engage in planning that results in an articulated curriculum based on standards and to design instructional materials that integrate content from other fields of study. This standard should be completed while demonstrating sensitivity to cultural and ethnic diversity, special needs, interests, abilities, and gender issues.

The next standard focuses on "Instructional Strategies." It addresses the ability of the teacher candidate to use a variety of effective teaching practices that enhance and extend learning about technology. The program should prepare a teacher candidate who can base instruction on contemporary teaching strategies that are consistent with the *Standards for Technological Literacy* (ITEA, 2000). The candidate should be able to apply principles of learning to the delivery of instruction and compare a variety of instructional strategies to maximize student learning. The teacher candidate should be able to use and demonstrate appropriate tools, materials, equipment, and processes to enhance student learning. He or she should be able to design and utilize a variety of student assessment strategies appropriate for the different instructional strategies. Finally, the candidate should be able to refine instructional strategies in order to improve teaching—learning by using self-reflection, student outcomes data, and other assessment techniques.

The eighth standard, "Learning Environment," is concerned with the ability of the teacher candidate to create and manage a learning environment that promotes technological literacy. The program should prepare teacher candidates who can design rich learning environments that provide for varied educational experiences in the technology classroom and laboratory. The learning environment should encourage, motivate, and support student learning, innovation, design, and risk taking. The teacher candidate should be able to create learning environments that are adaptable for the future. The candidate must exhibit safe practices and be able to design, manage, and maintain a physically safe technology education learning environment.

Obviously, the most important element in learning is the student. The ninth standard, which is titled "Student Learning and Diversity," focuses on the ability of the program to prepare teacher candidates who understand the needs of students as learners and how diversity affects learning. The program must prepare candidates who can design technology experiences for students of different ethnicities, socioeconomic backgrounds, genders, ages, interests, and exceptionalities. The candidate must be able to identify how students learn about technology most effectively by integrating current research about experiential and differentiated learning. The candidate should develop productive relationships with students so that they can become active learners.

The 10th ITEA/CTTE SPA standard is titled "Professional Growth." This standard encourages technology education programs to prepare teacher candidates who demonstrate an informed and continuously updated knowledge base about technological systems. Teacher candidates should be able to develop a professional development plan for their own future education and

Conducting Program Assessments

training. They should also be able to apply various marketing principles to promote technology education and the study of technology and to promote student organizations like the Technology Student Association (TSA). Candidates should collaborate with others to promote professional growth activities. They should actively participate in local, state, and national professional organizations. Finally, candidates should reflect upon their teaching to improve and enhance student learning.

For more information and specific details and benchmarks concerning the subject-matter standards (1–5), readers can refer to *Standards for Technological Literacy: Content for the Study of Technology* (ITEA, 2000). In addition, the document entitled *Accreditation Report: Five-Year Review of Revised Standards, Initial Programs in Technology Teacher Education,* (ITEA/CTTE/NCATE, 2004) contains "knowledge," "performance," and "disposition" indicators and appropriate rubrics for each of the 10 standards to help with program assessment. This document can be found on the CTTE Web site (www. ctteonline.org). Finally, the document entitled *Advancing Excellence in Technological Literacy: Student Assessment, Professional Development, and Program Standards* (ITEA, 2003) contains additional background knowledge about pedagogy standards 6 through 10.

ADDITIONAL PROGRAM ASSESSMENT COMPONENTS AND STRATEGIES

Advisory Committees

An advisory committee is an organized group of people who serve as a link between the college and the community. An advisory committee can help develop and refine the way the technology education program is designed and delivered. It can serve as both an advocacy group and a program improvement group.

Committee members should be selected for their thorough knowledge of the standards and the mission of the technology education program. Members are usually selected or recommended through a process established by the college department in charge of delivering technology education. In addition to representatives from technology education, members could also represent other areas of education (such as mathematics and science), the business and industry communities, the student body, alumni, governmental agencies, parents, and/or community agencies. Committee membership should be limited to a designated term length, and a staggered term system should be used so that the committee maintains an active and diverse perspective on the program. Committee membership usually runs from one to three years, although some members may remain active for extended periods because of their position or responsibilities.

Advisory committees should meet formally two or three times each year to review and discuss issues and concerns of the technology education program. The agenda may be set by the department faculty, department chair, committee chair, committee members, or a combination of these. The committee should establish a plan of work for each year. The advisory committee can make recommendations concerning areas such as new courses; existing course content and objectives; selection and evaluation of instructional materials and equipment; and the design, use, and updating of the facilities.

Advisory committees should examine program assessment data that are already collected by the technology education department or education unit so that they can make informed recommendations. This data could include course enrollments, number of program completer, state licensure test results, teacher candidate electronic portfolios, work samples, exit surveys, accreditation reports, and other forms of information.

Faculty Qualifications

Within the context of program evaluation, a key component is the assessment of faculty qualifications. It is obvious that the quality of the technology education program depends on the dedication, expertise, and efforts of faculty members within the department that delivers technology teacher education.

While there are no formal certification requirements for college faculty members, it is critical that college professors have extensive educational experience and training in the area of technology education. Most universities and colleges require or strongly recommend that professors hold master's or doctoral degrees in technology education or a closely related field. If the institution is a research-based institution, a doctorate is essential. In some state teacher colleges, a master's degree in technology education may be sufficient. In addition, some states also require that teacher education faculty hold licensure in the area in which they prepare teachers.

Faculty Professional Development

Just as the classroom teacher must participate in local and state professional development activities, college professors should participate in state and nationally sponsored workshops, conferences, and institutes. There are many national professional organizations to which technology educators can belong including the ITEA, the American Education Research Association (AERA), the Association for Career and Technical Education (ACTE), the Association of Supervision and Curriculum Development (ASCD), and others. States also have corresponding professional associations that conduct their own workshops and conferences that college faculty members can attend.

Involvement with professional development schools (PDSs) is an important professional development activity not only for the faculty and staff of the professional development school but also for the college faculty members who are involved with this activity. The activity keeps the college professor up-to-date about what is happening in the classrooms of today and the changes within public school teaching.

It is important that technology education programs have full-time faculty members whose primary emphasis is technology education. With the reduction in numbers of technology teacher education programs across the nation, declining student enrollments in technology education programs, and the consolidation of teacher education units on campuses, there seems to be a tendency to deliver technology education programs using part-time faculty members, adjunct faculty members, and faculty members outside the technology education field. In order to maintain the integrity of the technology education content area, programs should employ faculty with expertise in the area.

Faculty Evaluation

Faculty evaluation is also typically a component of the program assessment process. Each college has its own faculty evaluation system which usually involves both formative and summative assessments and is tied to a merit pay system. Most systems are broken down into teaching, research and scholarly activities, and service activities. Teaching is evaluated by such data as peer reviews, student course evaluations, and student grades and achievements. Research and scholarly activities include presentations at local, state, and national conferences, professional journal articles, books published, curriculum development and program design efforts, serving on graduate student committees, and involvement in grant writing as a principal or co-principal investigator. Service activities involve such things as participating on university, state, or national committees; reviewing grant proposals; hosting or judging student competitions; and serving as an officer in a local, state, or national professional organization.

According to Lunenburg and Ornstein (2004) and Herbert (2001), teacher electronic portfolios that contain "artifacts" are becoming more popular as a way to collect data on college faculty accomplishments. Electronic portfolios can convey a rich portrayal of teacher performance, including live-stream classroom activities, lesson plans, writing samples, reflective journals, and a variety of data such as student course ratings and grades.

ASSESSING PHYSICAL COMPONENTS OF TECHNOLOGY EDUCATION PROGRAMS

The assessment of facilities, equipment, materials, and instructional technologies is important to the overall assessment of the technology education program. With the change of content standards and more emphasis on design, problem solving, engineering, and technological systems, the technology education laboratory of today requires a different set of evaluation criteria than the industrial arts shops of the past.

The overall question to ask is whether the facilities, space, and equipment devoted to the technology education program are adequate for achieving the standards and outcomes specified for the technology education program. Assessing facility resources in today's technology classroom and laboratory would focus on the appropriateness of the equipment and materials available; whether adequate provisions for room temperature, humidity, exhaust, ventilation, acoustics, and illumination are present; if the lab can accommodate special needs students; aesthetic details such as cleanliness, layout, and design; and other features such as storage space and data network capacity.

Safety is another concern when assessing technology education programs. A complete assessment of the physical components of a technology education program would examine the extent to which safety provisions are in place, including safety instruction, safety audits, protective equipment, and maintenance records (Maryland State Department of Education, 1995).

CONCLUSION

This chapter addressed ideas and suggestions on how to perform assessments of technology education programs. To begin, 10 assessment principles were presented to lay the foundation for the program assessment process. These 10 principles should be used as a knowledge base of information and as a discussion guide for faculty to help foster and perform appropriate and meaningful program assessments in technology education.

There are many assessment techniques and measures used to collect data during the program assessment process. Some of the more popular measures include student portfolios, state licensure test scores, student GPA, alumni surveys, exit interviews, teacher work samples, and focus groups.

There are several accreditation and assessment organizations that can help guide the process of program assessment in technology education. These organizations include NCATE, NBPTS, CTTE, regional accreditation boards, and state departments of education.

Assessing the instructional content of technology education is of vital importance within the context of program assessment. It is important that curriculum design and assessment are consistent with the *Standards for Technological Literacy: Content for the Study of Technology* (ITEA, 2000). Program assessment in technology education would also not be complete without a thorough assessment of effective teaching practices. Assessment of teacher candidates typically occurs in the five areas of curriculum development, instructional strategies, learning environments, student learning and diversity, and professional growth.

Other programmatic issues covered in this chapter included using advisory committees, assessing faculty qualifications and performance, and faculty professional development. Advisory committees can be used for program improvement and advocacy purposes, and they can offer input into the program assessment process. The assessment of the physical components of a technology education program, such as facilities, equipment, materials, and safety, is also important.

By systematically applying the principles, strategies, and tools described here, program assessment will become part of an ongoing process of program improvement. In an age when program accountability and quality are increasingly important, technology education programs must adopt a proactive system of program review.

DISCUSSION QUESTIONS

1. In what ways could application of the 10 assessment principles described here help in designing and carrying out a program assessment?
2. What data collection strategies do you think would be most useful for program assessments in which program improvement is the key goal?
3. What organizations are involved in technology education program assessment, and how do they function?
4. What are the standards for assessing instructional content in technology education programs?
5. What are the standards for assessing effective teaching practices in technology education programs?
6. What are some contextual issues and areas that should be addressed when assessing technology education programs?

REFERENCES

Aaker, D. A. (1992). *Developing business strategies.* Berkeley, CA: University of California.

Banta, T. W., H. P. Lund, K. E. Black, and F. W. Oblander (1996). *Assessment in practice.* San Francisco, CA: Jossey-Bass Publishers.

Herbert, E. (2001). *The power of portfolios.* San Francisco, CA: Jossey-Bass Publishers.

International Technology Education Association (2000). *Standards for technological literacy: Content for the study of technology.* Reston, VA: International Technology Education Association.

——— (2003). *Advancing excellence in technological literacy: Student assessment, professional development, and program standards.* Reston, VA: International Technology Education Association.

ITEA/CTTE/NCATE (2004). *Accreditation report, five-year review of revised standards: Initial programs in technology teacher education.* Washington, DC: ITEA/CTTE/NCATE.

Lunenburg, F. C., and A. C. Ornstein (2003). *Educational administration: concepts and practices.* Belmont, CA: Wadsworth Publishing Co.

Maryland State Department of Education (1995). *Quality indicators for technology education programs in Maryland.* Baltimore, MD: Maryland State Department of Education.

Middle States Commission on Higher Education (2006). *Characteristics of excellence in higher education.* Philadelphia, PA: Middle States Commission on Higher Education.

Mitchell, A., S. Allen, and P. Ehrenberg (2005). *Spotlight on schools of education: Institutional responses to NCATE standards 1 and 2.* Washington, DC: NCATE.

National Board for Professional Teaching Standards (2006). About us. http://www.nbpts.org. Arlington, VA: National Board for Professional Teaching Standards (accessed January 20, 2006).

National Council for Accreditation of Teacher Education (2006). *Professional standards for the accreditation of schools, colleges, and departments of education.* Washington, DC: National Council for Accreditation of Teacher Education.

Southern Regional Education Board (2006). *From goals to results: Improving education system accountability.* Atlanta, GA: Southern Regional Education Board.

Assessing Teacher Readiness and Teaching Performance

Chapter 14

Kurt R. Helgeson
St. Cloud State University

CHAPTER OVERVIEW

Universities that prepare teachers have seen an increased demand for accountability that includes the need to provide high-quality teacher candidates and to demonstrate their effectiveness. School districts must also demonstrate and document the effectiveness of their teachers. This demand is not likely to go away or decrease in the foreseeable future. Therefore, it is important for educators at all levels to understand the role that assessment can play in improving both preservice and in-service teachers. There are a variety of approaches used in the technology education profession to assess teacher candidates prior to their graduation from college as well as after they become teachers in the field. Examples of common approaches will be examined in this chapter.

This chapter has been divided into two sections. The first section will look at assessment used during undergraduate preparation of technology education teacher candidates. The focus will be on how assessments are used to determine whether candidates are gaining the knowledge needed to prepare them for teaching. A significant part of the focus of the assessment at this level is on the performance of the teacher candidate during student teaching. Also discussed will be the various standardized tests and assessments used by the education profession. The second section will focus on assessment of existing technology education teachers. Most assessments of practicing teachers are part of the tenure and/or an annual review process, which may not address the need for assessing as a means of improving professional practice.

The goals of this chapter are to provide the reader
- information to develop an understanding of the role of assessment in teacher education programs;
- examples of how different states and accreditation organizations are using assessment to ensure the preparation of future teachers;
- information and examples of how similar assessments used in teacher education programs are being used for practicing teachers in K–12 schools; and
- an understanding of the important role assessment plays in improving all phases of teacher preparation and teaching practice.

ASSESSMENT TECHNIQUES FOR TEACHER CANDIDATE READINESS

There are a variety of general assessment techniques used throughout the courses included in undergraduate programs that are designed to determine if candidates are learning the desired content and skills. This section of the chapter will focus on specific techniques used for assessing students' progress toward development into teachers (referred to in this chapter as teacher candidates). In the past, such assessments simply looked at whether students were passing courses. As long as students were passing the courses and maintaining the required grade point average, they continued along in the program working toward student teaching and graduation.

This approach to assessment often ended with teacher candidates reaching student teaching only to find they did not enjoy teaching or did not have the skills to be effective teachers. The latter problem might have been a result of not gaining adequate technical preparation, not developing an understanding of pedagogy, or not having the dispositions to be a teacher—or possibly a combination of all three of these factors. Several assessment techniques have been introduced to many teacher preparation programs to try to ensure that all teacher candidates are well prepared for their roles as teachers, including the techniques used specifically by the technology education profession to prepare future technology education teachers.

UNIVERSITY MOVEMENT TOWARD ASSESSMENT

While the purpose of assessing student teachers is to determine if they should be recommended for licensure, assessment throughout the earlier undergraduate years does not have such a clear focus. Most of the movement toward incorporating assessment throughout the program has been stimulated by the accreditation process. The National Council for Accreditation of Teacher Education (NCATE) has partnerships with all 50 states as well as the District of Columbia and Puerto Rico. In 17 of the states, all public institutions that prepare teachers are accredited by NCATE.

As of 2006, there were 71 technology teacher education programs (International Technology Education Association [ITEA], n.d.) in the United States, 39 of which were nationally recognized by NCATE. Not all states require programs to be accredited by NCATE, but instead use some form of a state certification process. However, according to NCATE (2006), half of all states have either adopted or adapted the NCATE unit standards for use in their state approval processes, and half of all states rely solely on the NCATE accreditation process for approval of teacher education programs. In many cases, states use the specialty area standards developed by the ITEA and the Council on Technology Teacher Education (CTTE) as the basis for review of technology teacher education programs.

A primary advantage of going through the accreditation process is the focus on looking at the teacher candidate throughout the program by gathering documented evidence. The downside is that it is not a quick and simple process to develop an appropriate assessment plan, and this data collection process is ongoing, adding to teacher educator workloads.

One example of this move toward expanded assessment of teacher candidates can be found at Eastern Michigan University, where collection of data on teacher candidate outcomes has been mandated since 1993. All departments and programs are required to address four key assessment questions:

1. What are the key skills and concepts that all students should master before they complete the program?

Assessing Teacher Readiness and Teaching Performance

2. How will students be assessed to determine if the stated skills and concepts have been mastered?
3. What evidence is there that the skills and concepts have been mastered?
4. In cases where the identified skills and concepts have not been mastered, how will the program be revised to better prepare students to master the stated skills and concepts? (Bennion, Harris, and Work, 2005, p. 37)

In addition to NCATE, several organizations have developed standards for assessing teacher candidates, including the Interstate New Teacher Assessment and Support Consortium (INTASC) and the International Society for Technology in Education (ISTE), both of which will be discussed below. Different states have adopted a variety of approaches to assessing preparation of teacher candidates. What is common among all is the focus on accountability and the need to demonstrate student *outcomes,* as opposed to identifying the inputs (e.g., listing the required courses) on the assumption that desired outcomes will be or have been met, given the right inputs.

INTASC

INTASC is a consortium of state education agencies, higher education institutions, and national educational organizations dedicated to the reform of education, licensing, and ongoing professional development of teachers. The basic premise of the INTASC standards is that an effective teacher can integrate content knowledge with pedagogical understanding to ensure that all students learn. These standards examine knowledge, performance, and dispositions as three levels of teacher candidate performance, regardless of the discipline. There are 10 areas in which teachers need to be prepared to meet these standards. They include content knowledge, how students learn, student diversity in background and learning styles, instructional strategies, student motivation for positive learning environment, knowledge of communication techniques, ability to plan instruction, assessment, reflective practice, and communication within and outside the school (Interstate New Teacher Assessment and Support Consortium [INTASC], 2005). These standards are used to develop assessments for preservice teachers. For example, at St. Cloud State University, all of the field experience assessments, including student teaching evaluations, are based on the INTASC standards. The data collected from the assessments are used both for candidate assessment and program assessment. All of the evaluations are compiled by content area and provided to the respective departments for program assessment purposes.

After INTASC released its model core standards and achieved some consensus on what beginning teachers should know and be able to do, states turned their attention to how they might assess that knowledge and skill. Stakeholders determined that all teacher licensing exams should be standards-based; that a single licensing test would be inadequate since it would not provide enough evidence of a candidate's capabilities for a permanent teaching license; and that states should not only assess what a candidate knows, but whether or not the candidate can teach what he or she knows. The assessment method most commonly used to document that the candidate can teach is a teacher portfolio, which is a collection of artifacts that tell the story of a candidate's teaching as it develops over a period of time. The artifacts serve as evidence that demonstrates whether a teacher meets or exceeds the INTASC standards for beginning teachers (2005).

International Society for Technology in Education

The International Society for Technology in Education (ISTE) is a professional standards organization that focuses on the use of information technologies in educational settings. ISTE has developed three sets of standards known collectively as the National Educational Technology Standards (NETS): for students (NETS-S), teachers (NETS-T), and administrators (NETS-A). Most states have incorporated the NETS-T into their requirements for teacher education programs. As of May 2004, 38 states had adopted or had referenced the NETS-T standards for teachers, and 49 states had adopted or referenced the standards for at least one of the three groups (International Society for Technology in Education [ISTE], 2005).

NETS-T focuses on the capacity of teacher candidates to apply information technologies in educational settings. There are six standards categories, with performance indicators for each that provide specific outcomes to be measured when developing a set of assessment tools. The NETS-T categories include:

1. Technology operations and concepts;
2. Planning and designing learning environments and experiences;
3. Teaching, learning, and the curriculum;
4. Assessment and evaluation;
5. Productivity and professional practice; and
6. Social, ethical, legal, and human issues. (ISTE, 2005)

There is considerable overlap between the NETS-T and the INTASC standards. Both reflect on the core components of good teaching. The other commonality is the focus on performance. In this regard they differ from the Educational Testing Service's widely used Praxis tests (discussed in another section), which measure content knowledge and disciplinary understanding in a written test format. The NETS-T and INTASC standards require preparation of a portfolio that documents competent performance on the part of the teacher candidate. These portfolios are typically reviewed by faculty supervisors within the teacher education program and are retained to serve as critical evidence during accreditation review cycles such as NCATE.

TECHNOLOGY EDUCATION MOVEMENT TOWARD ASSESSMENT

The Council on Technology Teacher Education (CTTE) has aligned its program accreditation standards with the *Standards for Technological Literacy: Content for the Study of Technology* (ITEA, 2000). The ITEA/CTTE standards were approved by NCATE in Fall, 2003 (ITEA/CTTE, 2003). As the CTTE report to NCATE noted:

> In the field of technology teacher education, accreditation has taken on a more important role. In the past 15 years, guidelines for accreditation have been implemented into many universities.... With the help of the standards, the ITEA/CTTE/NCATE curriculum standards have been completely rewritten as performance-based standards and will continue to guide the technology teacher education profession. While these are appropriate for the discipline of Technology Education, many of the universities

will also have a campus wide accreditation, such as North Central Association. This may require additional or different data to be collected by the department. (ITEA/CTTE, 2003, p. 1)

A Sample Assessment Plan for ITEA/CTTE/NCATE Accreditation

Figure 1 provides an example of an assessment plan used by the Technology Education Program at St. Cloud State University (SCSU) in preparation for the ITEA/CTTE/NCATE accreditation review process. This information was developed to identify how the department was going to assess its teacher candidates at a variety of transition points. The department has identified these as critical assessment points to determine if a teacher candidate should move forward within the program. If it is determined that the teacher candidate is not ready to move forward, a variety of steps might be taken. These could include some type of remediation, career guidance for a change of majors, or removal from the program. Each department at SCSU is responsible for identifying its own transition points and sharing those with the College of Education. Many of the transition points are the same across departments, including admission to teacher education and student teaching.

Another transition point in the technology education program is completion of an introductory course in the department that is required of all majors. For technology education majors there are three assessed tasks within this introductory course. These include preparing a philosophy paper and a portfolio outline, as well as completion of the major application. The assessments are then scored with rubrics developed to help make scoring consistent among different faculty. The assessment plan stipulates what data is to be collected to document candidate performance at this transition point and what scores are to be reported to the College of Education. The final steps in the assessment plan address how the assessment will be used for program improvement. This section is designed to make sure that data collected from the assessments are reviewed and used by the department to improve the program for teacher preparation, as well as what the next steps are in this process. The SCSU example in Figure 1 is taken from the initial plan for preparing for an NCATE visit in 2007.

Therefore, many of the steps identified in the plan are already complete (as indicated by the due date for the work to be completed).

Figure 1. The SCSU assessment plan for technology education.

CANDIDATE/PROGRAM/UNIT ASSESSMENT PLAN
Environmental & Technological Studies Department
St. Cloud State University

Phase I: Transition Point 3: Completion of ETS156: Introduction to ETS.
Phase II: Identification of Major Assessment(s) at Transition Point:
    a. Philosophy of Education paper.
    b. Professional characteristic of a teacher paper.
    c. Successful completion of ETS156 or equivalent with a C or higher.
Phase III: Development of Rubrics for Transition Points
    a. Which of the major assessments have a rubric developed?
      -None
    b. Which of the major assessments need a rubric developed?
      -Assessment g and Assessment h
    c. Please attach a copy of the rubric for this transition point.
Phase IV: Data Collection
    a. What type of data on candidate performance can be collected and reported at this transition point?
      1. Number of teacher candidates completing the course.
      2. Statistical data on the scores on assessments g and h.
      3. Number of candidates requiring remediation on assessment g or h.
Phase V: Program Improvement
    a. How will this data assist in improving our program?
      1. It will help identify potential disposition problems with candidates earlier in the program. Candidates will have a better understanding of the characteristics of a teacher and a technology education philosophy at the start of their studies.
      2. Department and major assessment test will be administered to all seniors projects (required course) classes. Results from the test will cover all of the core course in the department. The results of the tests will be summarized and discussed at department meetings.

NEXT STEPS: *Phases 1-3 must be completed by December 1st.

Task	Person Responsible	Date Due
Rubric for Assessment g	Dr. Kurt R. Helgeson	11/15/03
Rubric for Assessment h	Dr. Kurt R. Helgeson	11/15/03

There are a total of seven transition points within the technology education program. Four of these transition points are the same regardless of the discipline of the teacher candidate, and all candidates must pass the assessment to continue in their respective programs. These are administered by the College of Education (transition points 1, 2 6, and 7). The transition points include

1. Education 300: Introduction to Teaching
2. Admission to the College of Education
3. Successful Completion of ETS 156: Introduction to Environmental & Technological Studies

4. Admission to Environmental & Technological Studies Department: Technology Education major
5. Completion of Technology Education major courses
6. Professional Digital Portfolio (see Figure 2)
7. Student teaching evaluations
8. Follow-up of teacher education candidates

Figure 2. Rubric for assessment of senior professional portfolio.

Portfolio Rubric

Evaluation Area	Poor	Average	Excellent
Content To what extent does the portfolio demonstrate the knowledge and experiences related to the intended career? Provide information on education, work experiences, professional organization, and other unique qualifications.	0–10 Some content but does not address all aspects of a career.	11–20 Most of the content is in the portfolio but lacks detail in the content.	21–30 Complete detail on a variety of content areas describing the experience & qualification.
Presentation What is the quality and completeness of the presentation of the materials? Is the organization logical and the navigation easy to understand and use?	0–10 Presentation and organization are weak. Navigation is difficult.	11–20 The presentation is adequate but could be improved.	21–30 Extremely well organized presentation.
Completeness Does the portfolio contain all of the information necessary to evaluate the qualifications to meet a variety of positions in the field of study?	0–10 Missing information that would be expected in a portfolio.	11–20 Lacks detail of experiences and education.	21–30 Some content but does not address all aspects of a career.
Writing Is the writing representative of a college graduate? Does the portfolio show a variety of writing samples, from technical to descriptive?	0–10 Limited demonstration of writing ability with spelling and/or grammatical errors.	11–20 Limited demonstration of writing ability with no spelling or grammatical errors.	21–30 Good demonstration of a variety of writing abilities and no spelling or grammatical errors.
Other Any other features that make the portfolio unique and communicate the qualifications to the employer.	0–10 Lacks any extra details to demonstrate knowledge and abilities.	11–20 Limited extra details to demonstrate knowledge and abilities.	21–30 Considerable extra details to demonstrate knowledge and abilities.

Data is collected and summarized for each of the transition points. The individual departments collect some of the data, and other data, including student teaching assessments, is collected by the College of Education for all programs. In this way, programs can identify needed changes, problem areas experienced by students at each transition point, and other issues requiring attention by the program faculty. The data is also used for program reviews and accreditation reports.

ASSESSMENT TECHNIQUES FOR TEACHER CANDIDATE TEACHING PERFORMANCE

The Educational Testing Service (ETS) has developed what is perhaps the most widely used set of written assessments for documenting content and pedagogy knowledge of teachers. The ETS Pre-Professional Skills Assessments (PPST®) are a series of standardized assessments designed to measure basic skills in reading, writing, and mathematics at the first level, content knowledge at the second level, and knowledge of teaching practice at the third level. Collectively, these tests are referred to as the Praxis exams.

PRAXIS I

Colleges and universities use Praxis I® to evaluate individuals for entry into teacher education programs. The assessments are generally taken early in a candidate's college career. Praxis I is used to rate competency in three general areas: reading, writing, and mathematics. These assessments are available in paper-based or computer-based formats. The reading and mathematics components of Praxis I each consist of 40 multiple-choice questions with 60 minutes of testing time allowed. The Writing component consists of 38 multiple-choice questions and one essay question with two 30-minute sections of testing time (Educational Testing Service [ETS], n.d.).

PRAXIS II

Praxis II®: Subject Assessment tests measure content knowledge of specific subjects taught at the K–12 level and also include general and subject-specific teaching skills and knowledge. There are Subject Assessments and Specialty Area Tests, Multiple Subject Assessment Tests for Teachers (MSAT), Principles of Learning and Teaching (PLT) Tests, and Teaching Foundations Tests. Individuals entering the teaching profession take these tests as part of the teacher licensing and certification process required by 34 states. Praxis II is also used by many states as a requirement for alternative licensure (ETS, n.d.).

The Praxis II: Subject Assessments and Specialty Area Tests measure general and subject-specific teaching skills and knowledge drawn from a particular discipline. The Specialty Area Test for Technology Education consists of 120 multiple-choice test items. The Technology Education test includes five sections, with the approximate percentage of questions for each section estimated as follows:

I. Pedagogical and Professional Studies: 30%
II. Information and Communication Technologies: 20%
III. Construction Technologies: 12%
IV. Manufacturing Technologies: 18%
V. Energy/Power/Transportation Technologies: 20% (ETS, n.d.).

The Principles of Learning and Teaching (PLT) test measures general pedagogical knowledge at four grade bands: Early Childhood, K–6, 5–9, and 7–12. These tests use a case study approach and feature constructed-response and multiple-choice items. Finally, the Teaching Foundations tests are used to measure pedagogical knowledge in five areas: multi-subject (elementary), English, language arts, mathematics, science, and social science. These tests feature constructed-response and multiple-choice items.

PRAXIS III

Praxis III®: Classroom Performance Assessments comprise a system for assessing the skills of beginning teachers in classroom settings. ETS developed Praxis III for use in teacher licensing decisions made by states, or for use by local agencies empowered to license teachers. Under the guidelines that govern its use, Praxis III may not be used for the purpose of making employment decisions about teachers who are currently licensed (ETS, n.d.).

This direct classroom assessment recognizes the importance of the teaching context as well as the many diverse forms that excellent teaching can take. The Praxis III system utilizes a three-pronged method to assess the beginning teacher's evidence of teaching practice. This includes direct observation of classroom practice, review of documentation prepared by the teacher, and semi-structured interviews.

Praxis III is an assessment system that is comprised of three separate, yet strongly interconnected, components. Individually, each component is designed to augment the value of the assessment. Collectively, the system is aimed at gaining a thorough understanding of the teaching skills of a beginning teacher. Overall, Praxis III provides insights into pedagogical areas in which a teacher may benefit from additional development.

According to ETS (n.d.) the components include
- **Component 1:** Framework of knowledge and skills for a beginning teacher that assesses the teaching performance across all grade levels and content areas.
- **Component 2:** Instruments used by trained assessors to collect data, analyze, and score the teaching performance.
- **Component 3:** Training of assessors to facilitate consistent, accurate, and fair assessments of a beginning teacher (PRAXIS III Overview, ¶3).

Beginning teachers residing and planning to teach in states that require Praxis III as one of the criteria for teacher licensing decisions (currently only Arkansas and Ohio) have their teaching skills assessed in classroom settings by trained assessors. The Praxis III Classroom Performance Assessments consist of a framework of knowledge and skills for a beginning teacher and contain 19 assessment criteria in four interrelated domains. These domains embrace the teaching and learning experiences of the beginning teacher, including:
- organizing content knowledge for student learning (planning to teach).
- creating an environment for student learning (the classroom environment).
- teaching for student learning (instruction).
- teacher professionalism (professional responsibilities; ETS, n.d., Testing Format, ¶2).

ASSESSMENT TECHNIQUES FOR IN-SERVICE TEACHER PERFORMANCE

There are three assessment areas in which in-service teachers are typically involved: (a) assessment by teachers to determine the effectiveness of instruction, (b) assessment of program effectiveness, and (c) assessment of teacher performance. The first and second areas are discussed in detail in earlier chapters. Most of the focus in this section will be on the third area. A variety of organizations—including NCATE, discipline-focused associations (such as the CTTE), and all state departments of education—have developed approaches and rationales for assessing in-service teachers. The two primary motivations for districts to develop and administer assessment programs are mandated annual teacher performance reviews and promotion and tenure reviews.

It is often difficult to separate assessment processes that are designed to focus on how to improve teaching from evaluations for purposes of performance, promotion, and/or tenure review. This is especially true in the current culture of accountability and demand for highly qualified teachers. According to the National Commission on Teaching and America's Future (2004), "The bipartisan passage of the No Child Left Behind [NCLB] Act of 2001 was an expression of national will. Recognizing that every American family deserves public schools that work, NCLB pledges highly qualified teachers in every classroom by the 2005–06 school year" (p. 4). Assessment plays an important role in the development, and designation, of highly qualified teachers. The Commission's report also notes:

> Successful schools are assessment centered. Teachers who are proficient in the use of well-designed assessment tools and strategies make learner-centered instruction possible. Sound assessment approaches provide continuous feedback that helps both students and teachers monitor learning while it is in progress. Revisions in the learning activities can be made as needed, and extra effort or new strategies can be tried before it's too late. (National Commission on Teaching and America's Future, 2004, p. 16)

It is particularly important for teachers in the field of technology education to be experienced in the use of diverse assessment techniques. The more commonly used assessment techniques generally don't fit with the hands-on activities prevalent in technology education. In many cases, assessment strategies need to be refined for best results. For example, a common assessment strategy is to evaluate a major project at the end of a course. By that time, it is often too late for adjustments to be made by the teacher and student. More frequent and less formal assessments used throughout the duration of the course would be a more effective approach.

A FRAMEWORK FOR ASSESSING TEACHER EFFECTIVENESS

Based in part on work done by the Educational Testing Service, Danielson (1996) published *Enhancing Professional Practice: A Framework for Teaching*. The importance of this work is that it parallels efforts made by other professions (such as medicine, law, or business) that over the years have developed their own expertise and procedures for certifying or licensing practitioners. Such efforts, when applied to education, can lead to the public's guarantee that teachers, like members of other professions, hold themselves to the highest standards. According to Danielson (1996), "a teacher makes over 3,000 non-trivial decisions daily... combining the skills of business management, human relations, and theater arts" (p. 2). This process is difficult at best for the beginning teacher. The role of assessment through the framework is to provide teachers from all experience levels and disciplines a means of communicating about excellence in teaching.

Danielson (1996) has identified "components of professional practice" for systematically looking at all aspects of teaching. These components are divided into four domains:
- Domain 1: Planning and Preparation
- Domain 2: The Classroom Environment
- Domain 3: Instruction
- Domain 4: Professional Responsibility (p. 4)

The domains are further broken down into several components. This model has provided the framework for many school districts on which to develop their teacher evaluation systems, and their mentoring systems for new teachers.

The framework for professional practice has many different uses. It provides a road map for novices by giving them a method for organizing what they need to do to prepare for teaching. It provides a structure for focusing improvement for either the veteran or novice teacher. Based on the framework, participants can conduct conversations about where to focus improvement efforts within the context of shared definitions and values, regardless of discipline. Finally, the framework provides a method to communicate about teaching to the larger community. What teachers do on a daily basis is generally a mystery to the general public. The four domains and their sub-components break down the teaching process into areas that are easy to understand. Planning and professional responsibilities are generally not seen when observing the classroom, but are critical to effective teaching. The framework provides the context for others to understand that activities within all four domains of teaching are taking place.

Assessing Readiness of Alternative Route Teachers

There has been considerable discussion in the past few years about alternative licensure. According to the NCATE (2006) Web site, there are over 130 institutions in the United States that offer alternative routes to licensure, including alternative licensure programs in 45 states. The requirements for the alternative license are as varied as the states and institutions themselves. Each state has different requirements which also vary by content area. These requirements range from no college classes required to additional college course work, documentation of expertise, and extensive supervision. The alternative license options generally feature on-the-job training for the how-to-teach component. The assumption is that the alternative route teacher has the necessary content knowledge and only needs to learn pedagogy.

In a study on alternative certification, Hoepfl (2003) found that the majority of teachers receiving alternative certification had previously taught in another field. The next most common background was a non-teaching technical background, followed by those with business/industry experience. In the study, which surveyed state supervisors, 71% of the respondents believed that the teachers with alternative certification were adequately prepared, while 19% felt they were not adequately prepared.

This has a great impact on the field of technology education because it has been identified as a critical shortage discipline. In the Hoepfl (2003) study, the top five states combined had nearly 600 unfilled technology education teaching positions. Florida alone had 150 unfilled positions. These positions are sometimes filled with emergency hires who will then have to work toward certification. Candidates for the permanent license are required to hold some type of bachelor's degree; beyond that, each case is typically reviewed individually to determine what other requirements are required for certification. The Praxis II or comparable state exam is often the final and determining factor in receiving a permanent teaching license (Hoepfl, 2003).

CONCLUSION

As John Goodlad (2004), through his work at the National Network for Educational Renewal, has indicated, we cannot have good schools without good teachers, and we cannot have good teachers without good schools. In order to have both good teachers and good schools, there is a need for the whole system to work together, and assessment is the key to this improvement.

DISCUSSION QUESTIONS

1. What are examples of assessment at the program level for technology education teacher candidates?
2. How does assessment differ for technology teacher education compared to teacher preparation programs in other disciplines?
3. What are the roles that assessment can play for in-service teachers? How does it differ from the role assessment plays for preservice teacher candidates?
4. What do you see occurring in the future regarding assessment of teachers?

REFERENCES

Bennion, D., M. Harris, and S. Work (2005). *Ten top barriers to assessment and how to overcome them.* Chicago, IL: The Higher Learning Commission.

Danielson, C. (1996). *Enhancing professional practice: A framework for teaching.* Alexandria, VA: Association for Supervision and Curriculum Development.

Educational Testing Service (ETS) (n.d.). Praxis test. http://www.ets.org/portal/site/ets/ (accessed June 15, 2005).

Goodlad, J. (2004). *A place called school.* Mission Hills, CA: Glencoe/McGraw-Hill.

Hoepfl, M. (2003). Alternative routes to certification of technology education teachers. *The Journal of Technology Studies* 37(2): 35–44.

International Society for Technology in Education (2005). *National education testing standards.* http://cnets.iste.org/ (accessed June 15, 2005).

International Technology Education Association (ITEA) (n.d.). *ITEA institutional members.* http://www.iteaconnect.org/Resources/institutionalmembers.htm (accessed September 28, 2006).

International Technology Education Association (ITEA) (2000). *Standards for technological literacy: Content for the study of technology.* Reston, VA: International Technology Education Association.

International Technology Education Association/Council on Technology Teacher Education (2003). *ITEA/CTTE/NCATE curriculum standards: Initial programs in technology teacher education.* Reston, VA: International Technology Education Association/Council on Technology Teacher Education.

Interstate New Teacher Assessment and Support Consortium (2005). *National educational technology standards.* http://www.ccsso.org/projects/ (accessed June 15, 2005).

National Commission on Teaching and America"s Future (2004). *What still matters most: Quality teaching in schools organized for success.* Washington, DC: National Commission on Teaching and America's Future.

National Council for Accreditation of Teacher Educators (n.d.). *NCATE and the states.* http://www.ncate.org/states/NCATEStates.asp?ch=95 (accessed September 28, 2006).

INDEX

A

accountability, 1, 4, 246–247. *See also* Vocational Competency Achievement Tracking System (VoCATS), North Carolina Department of Public Instruction (NCDPI)
ACCUPLACER tests, 143
achievement
    assessment implications of, 208–209
    in assessment square, 212–215
    framework of, 204–206
    knowledge types in definition of, 206–208
Advancing Excellence in Technological Literacy (International Technology Education Association), 4, 22, 88, 266
advisors, practicum, 258
advisory committees, 266–267
affective assessments, 93–94
affective domain, 237–238, 240
Aikenhead, G., 162
Albion, P., 165
Alexander, P. A., 205, 238
All Americans Project, 19
Almond, P., 93
Almond, R. G., 211
alternative assessment tools, 2, 65–86
    classroom-based, 74–79
    in design and technology (D&T, U. K.), 195–198
    educator advocacy of, 82–83
    large-scale case study of, 79–82
    overview of, 65–67
    scoring with, 67–73
    for students with disabilities, 93–95
    teaching licensure and, 287
    in trade and industrial education, 240–246
American Association for the Advancement of Science (AAAS), 22
American Association of Colleges for Teacher Education (AACTE), 259
American Education Research Association (AERA), 268
Americans with Disabilities Act (ADA) of 1990, 95–96
analysis. *See also* data analysis

analytic rubrics, 68, 70
Ancess, J., 83
ancillary skills, in assessments, 213
Anderson, J. R., 205, 206–207, 213, 217
Anderson, L., 34, 69–70, 238
Angrist, J. D., 168
Ankiewisc, P., 161
Anoka-Hennepin Schools, 1, 103
Appalachian State University, 1, 65
apprenticeships, 33
Archer, B., 187
assessments, 1–16
    issues in, 4–6
    purpose of, 6–8
    technology education areas for, 9–10
    timing of, 8
    value of, 10–14
    yearbook overview of, 1–4
    See also alternative assessment tools; teacher performance
assessments, communicating results of, 125–137
    formative, 125–128
    modifying instruction after, 133–135
    self-, 128–129
    summative, 129–132
assessments, conducting, 251–272
    advisory committees and, 266–267
    of faculty evaluation, 268–269
    of faculty professional development, 268
    of faculty qualifications, 267–268
    of instructional content, 262–264
    measures and techniques of, 256–259
    organizations for, 259–262
    principles of, 252–256
    of program physical components, 269–270
    of teacher candidates, 274
    of teaching skills, 264–266
Assessment of Performance in Design and Technology, The (Kimbell), 27
Assessment of Performance Unit, Goldsmiths College, University of London, 159, 162
Assessment of Performance Unit (APU), Department for Education and Science (U. K.), 185

291

Index

assessment square, 210–218
    achievement framework in, 212–215
    cognitive analysis in, 217–218
    components of, 210–212
    retrospective logical analysis in, 215–217
assessment triangle, 209
ASSET tests, 143
assistive technology software, 92
Association for Career and Technical Education (ACTE), 268
Association of Colleges and Schools, 246
Association of Supervision and Curriculum Development (ASCD), 268
Atkin, J. M., 222
atomism, in assessment, 162
"at risk" students, 142
attitudes, workplace, 237
"authentic" assessments, 93
automaticity, in procedural knowledge, 206
Ayala, C. C., 209–210

B

Bak, H. J., 159, 168
Bame, E. A., 160–161
Banta, T. W., 252
Baron, J. B., 220
Barton, K., 90
Baxter, G. P., 204, 211, 218
Becker, K., 161
Bell, B., 223
benchmarks, 74
Bennett, R. E., 217
Bennion, D., 276
Bernhardt, V. L., 117
Berrett, J. V., 3, 125
Bertrund, A., 38
"best-answer" approach, 42
best practices, teaching to standards as, 9
bias-free assessments, 90, 105
Black, K. E., 252
Black, P., 5, 219, 221–223
Blomgren, R. D., 169
Bloom, B. S., 34, 140
Bloom's taxonomy, 34. *See also* Revised Bloom's Taxonomy (RBT)
"blueprints," 38–39, 140

Bobbitt Nolen, S., 212, 221
Borich, G. D., 73
Boser, R. A., 161
Boster, J., 163–165
Brennan, R. L., 11
Breyer, F. J., 211
Bridgeford, N. J., 33
Brigden, C., 160
Brigham Young University, 125
Brophy, J. E., 127
Brown, J. S., 67, 207
Bruner, J. S., 188
Buchanan, R., 188, 192–193
Buehl, D., 128
Buffer, J. J., 92, 94
Burdette, P., 93
Burdge, M., 93
Burdin, J. L., 33
Burns, J. Z., 3, 233, 241, 244
Burnstein, R. A., 65
Bybee, R. W., 205

C

Cambre, M. A., 165
Campbell, J., 167, 171
capability, assessing, 26–27
capstone experiences, 258
Cardon, P. L., 2, 87
career and technical education (CTE), 79–82. *See also* Vocational Competency Achievement Tracking System (VoCATS), North Carolina Department of Public Instruction (NCDPI)
Carl D. Perkins Career and Technology Education Act of 1984, 79, 95, 97–98
Carl D. Perkins Vocational and Applied Technology Education Act of 1990, 97, 141, 246
Carlo, M. S., 206
Carlson, W. B., 186
Carter, K., 33
Cebula, J. P., 38
certification requirements, 3, 154
Chambers, B., 33
Champagne, A. B., 220
Chappuis, J. A., 107

Index

Chappuis, J. S., 107
cheating on tests, 52
checklists, evaluation, 68, 242–243
Chi, M. T. H., 205
Children's Attitudes Toward Technology Scale, The (CATS), 166
Chown, A., 134
Christensen, R., 165
Chudowsky, N., 5, 203, 205, 209–211, 218–219
Ciscero, C. A., 206
Civil Rights Act of 1974, 96
classroom-based assessments, 7, 33–64
    administering, 50–52
    analyzing, 52–57
    computers in, 58
    essay test items for, 45–48
    importance of, 34–36
    limitations of, 59
    matching test items for, 44–45
    multiple-choice test items for, 41–44
    objectives in, 38
    problem-based test items for, 49–50
    in science and mathematics, 219, 222
    short-answer test items for, 48–49
    test items for, 36–37
    true-false test items for, 39–41
    See also alternative assessment tools
"cleaning" data, 112–113
Coalition for Evidence-Based Policy, 10
Coffey, J. E., 222–223
cognitive analysis, 217–218
cognitive demands, in assessments, 213
cognitive dimension, of achievement, 205
cognitive domain, 237–238, 242
cognitive modeling, 187
collaboration, of students, 94
college-entrance exams, 7
commercially prepared assessments, 7
Committee on Information Technology Literacy, 167
communicating results of assessment. *See also* assessment, communicating results of
complexity of items, in assessments, 213
computer-based assessments, 4, 12, 58, 92
Computer Literacy and Computer Science Tests, 166

concept mapping, 66
conceptual analysis, in assessment square, 211
Concord Consortium, 27
conducting assessments. *See also* assessment, conducting
Conoley, J. C., 166
content dimension, of achievement, 205
content validity, 53–54, 89, 106–108
Cook, D., 165
cooperative assessments, 91, 245
Cope, B., 158–159
core content areas, 5, 10
Council of Chief State School Officers (CCSSO), 259
Council on Technology Teacher Education (CTTE)
    accreditation by, 278–281
    standards of, 260, 262, 264, 266, 275
    Yearbook of, 1
Cowie, B., 223
craft-based traditions, 181
Craft Design and Technology (CDT), 182
Crawford, L., 93
Cronbach, L. J., 211, 218
CTB/McGraw-Hill TestMate software, 140–142
Cuban, L., 168
curriculum standards, 67
Custer, R. L., 2, 17, 19, 26, 241

D

Dakers, J. R., 158
Daniels, H., 113
Danielson, C., 285–286
Darling-Hammond, L., 83
Dartmouth College, 19
data analysis, 103–124
    of classroom-based assessments, 52–57
    data gathering and, 109–111
    data warehouses for, 115–117
    group student, 114–115
    importance of, 103–104
    individual student, 113–114
    linking elements of, 117
    overview of, 13–14
    planning assessments for, 112–113
    purpose aligned with, 104–108

Index

reporting on, 119–121
statistical software for, 117–118
on STL, federal, and state mandates, 118–119
of Vocational Competency Achievement Tracking System (VoCATS), 149–153
Daugherty, M., 161
Day, G. F., 3, 251
declarative assessments, 6, 8
declarative knowledge, 106–107, 204, 206, 208
deficit models, 168–169
de Jong, T., 205
De Kleer, J., 207
de Klerk Wolters, F., 160
De Lisi, R., 130, 133
Department for Education and Employment Qualifications and Curriculum Authority (DfEE/QCA, U. K.), 183, 194–195, 198
Department for Education and Science (U. K.), 185, 190
design and technology (D&T, U. K.), 181–202
alternative performance assessment in, 195–198
curriculum and assessment in, 198–200
design process core in, 181–185
design thinking in, 185–187
large-scale assessments of, 162–163
learning power in, 187–191
performance assessment in, 191–195
Design & Technology Teachers' Professional Association (DATA), 195
de Vries, M., 19, 160–161
Digital Transformation: A Framework for ICT Literacy (Educational Testing Service), 158
disabilities, assessing students with, 2, 87–102
alternative tools for, 93–95
background on, 87–88
bias-free, 90
communicating need for, 98
importance of, 88
on laboratory safety, 91
legislation on, 95–98
modifying traditional tools of, 91–92
multi-factored, 89
validity in, 89–90
in Vocational Competency Achievement Tracking System (VoCATS), 142

Disciplinary Understanding and Attitudes Toward Technology (Siciliano and Maser), 169
disclosure statements, on assessments, 131
Discriminating Power ratings, in item analysis, 56
Doretz, D. M., 90
DTI Associates Inc., 234
Duchastel, P. C., 51
Duell, O. K., 41
Dugger, W., 19, 160–161, 165
Dunnell, P., 41
Duschl, R. A., 223

E

Earth First!, 164
Eastern Michigan University, 87, 275
Educational Testing Service (ETS)
Digital Transformation: A Framework for ICT Literacy publication of, 158
ICT Literacy Assessment of, 167
Large-scale assessments of, 157, 167–170
PRAXIS assessments of, 167, 282–284
Pre-Professional Skills Assessments (PPST) of, 282
teacher effectiveness and, 285
Education for All Handicapped Children Act (EAHCA) of 1975, 4, 87, 96
Edwards, E., 239
efficiency, testing and, 35
Eindhoven University of Technology (Netherlands), 19
Elder, A. D., 204
Ellsworth, R. A., 41
Ellwien, M. C., 33
Elshof, L., 164
empirical analysis, in assessment square, 211–212
"employability chart," 245
employer surveys, as assessment technique, 258–259
enabling skills, in assessments, 213
English proficiency, limited, 142
Enhancing Professional Practice: A Framework for Teaching (Danielson), 285
Environmental Values instrument (Kempton, Boster, and Hartley), 164

environments for assessment, simulated, 27
-portfolio system (E-scape, U. K.), 27
Ernst, J. V., 96
E-scape (U. K.), 27
Eshet-Alkalai, Y., 167
essay test items, 37, 45–48, 91–92
evaluation checklists, 68, 242–243
evidence-based educational change, 10
exit interviews, as assessment technique, 257
externally mandated assessments, 219

F

face validity, 89, 106, 108
faculty, assessments of, 267–269
Falk, B., 83
Farr, M., 205
Feltovich, T., 205
Feng, F., 165, 171
Fen Lin, M., 168
Ferguson-Hessler, M., 205
Fisher, M., 165
Fleming, M., 33
Fleming, R., 162
focus groups, as assessment technique, 258
formative assessments, 5–6, 125–128, 133, 223–224
Fourez, G., 168
Frantom, C., 166
Functional Information Technology Test (FITT), International Association for the Evaluation of Educational Achievement (IEA), 166
Furtak, E. M., 223
Future, modeling of, 189–190

G

Gallup polls, 165
Gao, X., 211, 218
Garden, R., 160
Garmire, E., 19, 170
Garrett, B., 93
Gelman, R., 204
Genova, P., 188
Gentner, D., 207
Georgia Department of Education, 239
Georgia State University, 233
Gerlach-Downie, S., 65

Ghiselli, E., 167
Gitomer, D. H., 223
Glaser, R., 5, 203–205, 209–211, 218–219
Glasgow, A., 93
Glass, C. R., 165
Gleser, G. C., 218
goal-directed play, design as, 188
Goh, D. S., 91, 92–93
Goldsmiths College, University of London, 19, 159, 162, 181, 195
Good, T. L., 127
Goodlad, J., 288
Gorman, M. E., 186
government testing, 25. *See also* National Assessment of Educational Progress (NAEP)
grade-point average of students, as assessment technique, 257
graduate exit surveys, as assessment technique, 258
graphic organizers, 66
Green, K., 166
Green, R., 162
Greeno, J. G., 204
Gronlund, N. E., 38
group-oriented assessments, 93–94
growth scores, 105, 109
Gullickson, A. R., 33
Guo, R. X., 3, 157, 166
Guryan, J., 168

H

Haertel, E., 213, 217–220
Haertel, G. D., 65
Hameed, A., 169
Hamilton, L. S., 205, 218
Harris, C., 76
Harris, M., 276
Harrison, C., 5
Hartley, J., 163–165
Hatch, L. D., 25
Hawisher, G., 170
Hayden, M., 169
Haynie, W. J., 2, 13, 33, 41, 43, 51–52, 57, 60
Heinssen, R. K., 165
Helberg, C., 104
Helgeson, K. R., 3, 273

Index

Herbert, E., 269
Heward, W. L., 89, 96–98
hierarchical linear modeling (HLM), 109
high-stakes assessments, 5, 7, 11
Hill, R. B., 234, 240
Hoepfl, M., 1, 2, 12–13, 41, 65, 79–80, 82, 148–149, 170, 287
Hoffman, E., 166
holism, in assessment, 162, 198
holistic rubrics, 68, 70, 78
Hollenbeck, K., 93
Hollinger, C., 89
Holly, P. J., 121
Howie, S. J., 159
How People Learn (National Research Council), 128
Hyde, A., 113
Hynds, P., 77–78

I

ICT Literacy Assessment (Educational Testing Service), 167
"if-then" rules, in procedural knowledge, 206
Illinois State University, 17, 19
Individual Education Plans (IEPs), 88, 91, 98
Individuals with Disabilities Education Act (IDEA) of 1975, 95–97
Individuals with Disabilities Education Act of 1990, 87
Industrial Designers Society of America (IDSA), 170
industry education. *See also* trade and industrial (T & I) education
inferential measurement, 171
informal formative assessment, 223–224
Initial Programs in Technology Teacher Education (International Technology Education Association, Council on Technology Teacher Evaluation, and NCATE), 266
in-service teachers. *See also* teacher performance
instructional management software, 140
instructional validity, 73
INTASC (Interstate New Teacher Assessment and Support Consortium), 276–278
interactive assessments, 92

international assessment approaches. *See also* design and technology (D&T, U. K.); science and mathematics assessment
International Association for the Evaluation of Educational Achievement (IEA), 166
International Society for Technology, 158
International Society for Technology in Education (ISTE), 276–278
International Technology Education Association (ITEA). *See also Advancing Excellence in Technological Literacy* (International Technology Education Association); *Measuring Progress: Assessing Students for Technological Literacy* (International Technology Education Association); *Standards for Technological Literacy* (STL, International Technology Education Association)
 accreditation by, 279–281
 on alternative assessment tools, 66
 on communicating assessment results, 130
 for faculty professional development, 268
 ITEA/Gallup poll of, 165
 National Academy of Engineering and, 18
 NCATE-approved standards of, 262–264
 on program assessment techniques, 259–260
 publications of, 4, 22, 28, 67
 rubric template of, 71
 on self-assessment, 128
 teacher education programs and, 275
 teaching skills standards of, 264–266
intervention strategies, 89
interviews, as assessment technique, 218, 257
Irwin, A., 159, 168
item analysis, of tests, 54–57
item openness, in assessments, 213

J

Janssen Reinen, I. A. M., 166
Johnson, D. W., 91
Johnson, L., 211
Johnson, R. T., 91
Jones, K., 82
Jones, M., 165
Judy, J. E., 205

K

Kahn, R., 159
Kalantzis, M., 158–159
Keirl, S., 158
Kellenberger, D., 165
Kelley, V., 162
Kellner, D., 159
Kelly, D. L., 159
Kelly, V., 186
Kemp, J. E., 234–235
Kempton, W., 163–165
Kendall, J. S., 160
Kenny, B., 170
Kim, J., 171
Kimbell, R., 3, 13, 19, 27, 162, 181, 186–187, 198–199
Kimeldorf, M. R., 87, 91
Kirkpatrick, H., 168
Klein, S., 205
Knezek, G. A., 165–166
Knight, L. A., 166
knowledge
    assessing, 9
    recall, 38
    task-related, 190–191
    in Third International Mathematics and Science Study (TIMSS), 208
    types of, 3, 204, 206–208
Kohn, A., 5, 11, 135
Kossar, K., 94
Krajcik, J., 76
Kramer, J. J., 166
Krathwohl, D. R., 140, 238
Kuforiji, P. O., 169
Kupermintz, H., 218

L

laboratory safety, 91
LaPorte, J. E., 96
LaPorte, T., 164
large-scale assessments, 3, 7, 157–179
    alternative, 79–82
    Blomgren *Understanding American Industry Test* as, 169
    British design and technology, 162–163
    of capability, 27
    classroom assessments versus, 222
    deficit models and, 168–169
    of Educational Testing Service (ETS), 167–170
    externally mandated, 219
    of International Association for the Evaluation of Educational Achievement (IEA), 166
    ITEA/Gallup poll as, 165
    of National Academy of Engineering (NAE), 170–171
    National Assessment of Education Progress (NAEP) as, 160
    of National Science Foundation (NSF), 163–165
    overview of, 157–159
    PATT instrument for, 160–161, 169
    Third International Mathematics and Science Study (TIMSS) as, 159–160
    of University of British Columbia, 166–167
Last, J., 134
Le, V. H., 218
Leach, J., 157
Learning Logs, for self-assessment, 128
Lederman, L. M., 65
Lee, C., 5
Lee, S., 168
legislation, on students with disabilities, 95–98
Lehmann, I. J., 33, 59
Lewis, R. B., 88
Lewis, T., 170
Lezotte, L. W., 11
Li, M., 204–205, 208–211, 213–214, 217
licensure exams, 7, 28, 257, 287. *See also* teacher performance
limited English proficient students, 142
Lindstrom, M., 1, 2, 103
Linn, R. L., 213
literacy, technological. See technological literacy
live performances, as alternative assessment tool, 66
Lizotte, D., 76
Lo, T. K., 161
Loepp, F. Z., 96

Index

logical analysis, 211, 215–217
Lund, H. P., 252
Lunenburg, 269

M

manual arts education, 33
Marshall, B., 5
Marshall, M., 160
Martin, M. O., 159
Martinez, M. E., 217
Marx, R., 76
Maryland State Department of Education, 270
Marzano, R. J., 12, 160
Maser, B., 169
Masia, B. B., 140
matching test items, 37, 44–45
mathematics assessment. *See also* science and mathematics assessment
Maunsaiyat, S., 161
Mayer, R., 207
McBee, J., 160–161
McDonald, M., 93
McLaughlin, C., 164
McLeod, S., 118
McLoughlin, J. A., 88
McNamara, K., 89
McNeill, K., 76
McTighe, J., 67, 108
Measuring Progress: Assessing Students for Technological Literacy (International Technology Education Association), 4, 67, 74, 158
Mehrens, W. A., 33, 59
Messick, S., 204, 209, 217
Metlay, D., 164
Michko, G. M., 168
Microsoft Excel statistical software, 118
Middle States Commission on Higher Education (MSCHE), 260–261
Miller, J., 163
Ming, Y. M., 161
Minitab statistical software, 117
Mislevy, R. J., 211
Missouri Department of Elementary and Secondary Education, 67
Miyashita, K., 166

models
  cognitive, 187
  deficit, 168–169
  of future, 189–190
  hierarchical linear (HLM), 109
  observation, in assessment square, 211
  self-assessment, 129
Morin, J., 134
Morrison, G. R., 234–235
Mullis, I. V. S., 220
multi-factored assessments, 89
multiple-choice test items, 37, 41–44, 92
Murphy, L. L., 166
Murray-Ward, M., 242
Myburgh, C., 161

N

Nagle, B., 77–78
Nanda, H., 218
National Academies, 2, 17–31
  National Academy of Engineering in, 18
  National Science Foundation in, 18–19
  performance assessment and, 26–27
  technological literacy assessment and, 19–25
    case for, 19–21
    framework for, 21–23
    improving, 28
    national agenda for, 24–25
  technology education teachers and, 29
National Academy of Engineering (NAE), 2, 17
  International Technology Education Association (ITEA) and, 18
  large-scale assessments of, 157, 170–171
  publications of, 20, 22
National Academy of Science, 17
National Assessment Governing Board (NAGB), 25, 29
National Assessment of Educational Progress (NAEP), 7, 22, 24, 160
National Association of State Directors of Teacher Education and Certification (NASDTEC), 259
National Board for Professional Teaching Standards (NBPTS), 261
National Center for Education Statistics, 88

National Commission on Teaching and
  America's Future, 285
National Council for Accreditation of Teacher
  Education (NCATE), 3, 259, 262, 275,
  279–281, 284
National Curriculum of England and Wales, 162
National Education Association (NEA), 259
*National Education Technology Standards for
  Students*, International Society for
  Technology, 158
*National Education Technology Standards
  (NETS)*, International Society for
  Technology in Education (ISTE), 277–278
National Network for Educational Renewal, 288
National Research Council (NRC), 2
  on classroom assessments, 222
  large-scale assessments of, 219–220
  in National Academies, 17–18
  publications of, 20, 22, 128
  traditional testing and, 65
National School Boards Association (NSBA),
  260
National Science Foundation (NSF), 18–19,
  28–29, 163–165, 170
Nation's Report Card, 22
New Technology Index, 166
Nitko, A. J., 33
No Child Left Behind (NCLB) Act of 2001, 4,
  20, 24, 95–96, 153, 155, 246, 285
Nolen, S., 212, 221
norm-based references, 157
North Carolina Department of Public
  Instruction (NCDPI), 79, 139, 145. *See also*
  Vocational Competency Achievement
  Tracking System (VoCATS), North Carolina
  Department of Public Instruction (NCDPI)
Northern Illinois University, 163
Nungester, R. J., 51
Nussbaum, E. M., 218

O

Oakes, J., 220
objectives, educational, 38
objective tests, 91–92
Oblander, F. W., 252
observation models, in assessment square, 211

Oderkirk, J., 165
open-source philosophy, 169
Organisation for Economic Co-operation and
  Development, 205
Orlansky, M. D., 89
Ornstein, A. C., 269
Orpwood, G., 160

P

Palmer, J. D., 161
Papanek, V., 188
parent-teacher conferences, 131
PATT instrument, 160–161, 169
Pearson, G., 2, 17, 165, 170
Pearson, R., 165
Peer tutors, 91
Pelgrum, W. J., 166
Pellegrino, J. W., 5, 9–10, 203, 205, 207, 209–211,
  218–219
performance-based assessments, 3–4
  alternative, 195–198
  data analysis of, 104
  in design and technology (D&T, U. K.),
    191–195
  National Academies and, 26–27
  National Assessment of Educational
    Progress (NAEP) as, 24
  North Carolina Department of Public
    Instruction (NCDPI) and, 81
  potential for, 12–13
*Performance Based Education Implementation
  Handbook* (Missouri Department of
  Elementary and Secondary Education), 67
performance lists, 68
performance products, as alternative assessment
  tool, 66
Peterson, R. E., 96
Petrina, S., 3, 157–158, 165, 168, 170–171
Pitfalls of Data Analysis (Helberg), 104
Pivot Tables, 118
placement rates, as assessment technique, 257
Plake, B. S., 166
play, design as, 188
Plomp, T., 166
Podemski, R. S., 88
politics of assessment, 4

= 299 =

Index

Popham, W. J., 1, 11, 90, 94
portfolios
    as assessment technique, 257–258
    electronic, 27
    of students with disabilities, 93–94
    for teacher evaluation, 133, 269
Power, P. W., 89, 92
practicum advisors, 258
Praxis assessments, Educational Testing Service (ETS), 7, 28, 167, 257, 282–284
predictive validity, 89
Pre-Professional Skills Assessments (PPST), Educational Testing Service (ETS), 282
preservice teachers. See teacher performance
probability, 110–111
"probe studies," 25
problem-based test items, 37, 49–50
problem solving, 188, 237–239, 242
procedural assessments, 6
procedural knowledge, 106, 204, 206
professional certification, 3, 154
Programme for International Student Assessment (PISA), Organisation for Economic Co-operation and Development, 205
Progress in International Reading Literacy Study (PIRLS), 170
proximal assessment tasks, 222
psychomotor skills, 93, 95, 237, 239
Public Opinion Laboratory, Northern Illinois University, 163
p-values (probability), 110–111

Q

Quinlan, A., 68–69, 71–72

R

Raat, J. H., 160
Raizen, S. A., 220
Rajaratnam, N., 218
randomization, selective, 127
rating scales, 243–244
reflective practitioners, teachers as, 133–135
Reflect/Reflect/Reflect, for self-assessment, 128
Reineke, R., 133

reliability
    of multiple-choice test items, 41
    of performance assessment data, 194–195
    of rubrics, 72–73
    of standardized tests, 167
    validity and, 53–54
response assessments, 91–92, 211
Revised Bloom's Taxonomy (RBT), 34–35, 40, 43, 69–70, 140
Richter, J., 169
Roberts, P., 187
Robinson, M., 170
Robitaille, D., 160
Rogers, G., 170
Rose, L., 165
Rosen, L., 166
Rosenthal, R., 127
Ross, S. M., 234–235
Roth, W. M., 168
Royer, J. M., 206
rubrics
    analytic, 68, 70
    development of, 75, 80–81
    in evaluating assessments, 209
    holistic, 68, 70, 78
    for portfolios, 281
    reliability of, 72–73
    for self-assessment, 129
    templates for, 71
    in trade and industrial education, 242
Ruiz-Primo, M. A., 3, 13, 203–205, 208–214, 217–218, 220, 223–224
Ryan, A., 162

S

Sadler, D. R., 222–223
safety, 91, 270
Sakamoto, T., 166
sample-based performance assessments, 24
Sanders, M. E., 96
Sarkees-Wircenski, M. D., 92–93, 95, 96–98
Saskatchewan Education, 162
Satchwell, R. E., 96
SAT exams, 7
Schaefer, K. M., 3, 233
schematic knowledge, 204, 207

Index

Schultz, S. E., 211, 213
Schwaller, A. E., 3, 251
science, technology, engineering, and mathematics (STEM) education, 5
science and mathematics assessment, 203–231
    achievement definition in, 204–209
    assessment square for evaluating, 210–218
        achievement framework in, 212–215
        cognitive analysis in, 217–218
        components of, 210–212
        retrospective logical analysis in, 215–217
    function and form in, 218–223
    informal formative, 223–224
Science Education for Public Understanding Program (SEPUP), University of California-Berkeley, 77–78
scope and sequence, in technology education, 145–149
scorecards, 244–245
scoring
    alternative assessment tools, 67–73
    in assessments, 211
    on growth, 105, 109
Scott, J. L., 66, 92–93, 95–98
Scott, M. L., 92, 94
Secondary Examinations Council (U. K.), 192
selective randomization, 127
self-assessment, 69, 128–129, 245
Selfe, C., 170
Self-Monitoring Approach to Reading and Thinking (SMART), for self-assessment, 128
Shavelson, R. J., 204–205, 208–211, 213–215, 217–218, 223
Shepard, L. A., 65, 74, 76–77, 208, 218, 220–221
short-answer test items, 37, 48–49, 92
Shown, T., 3, 139
Shumway, S. L., 3, 125
Siciliano, M., 77–78
Siciliano, P., 169
Siegal, M., 77–78
Sierra Club, 164
Silver, E., 203
simulated environments for assessment, 27
simulations, computer-based, 92
Singer, D., 188

Singer, J., 188
Sixteenth Mental Measurements Yearbook, The (Spies and Plake), 166
small-scale assessment, 171
Smith, J. K., 130, 133
Smith, L. F., 130, 133
Snow, R. E., 203, 217–218
Southern Regional Education Board (SREB), 260–261
Special Professional Association (SPA), Council on Technology Teacher Evaluation (CTTE), 260, 262, 264, 266
Spies, R. A., 166
SPSS statistical software, 117
St. Cloud State University, 251, 273, 276–277
Stables, K., 162, 186, 189
stakeholders, educational, 129
Standardized Test of Computer Literacy and Computer Anxiety Index (Conoley and Kramer), 166
standardized tests, 11
standards, 9, 67, 74, 76, 221
Standards for Technological Literacy (STL, International Technology Education Association), 4
    adoption of, 14
    as assessment framework, 22–23
    data analysis on, 107–108, 111, 118–119
    impact on instructional content of, 134, 262–264, 270
    NCATE approval of, 278
    North Carolina Department of Public Instruction (NCDPI) and, 147, 155
    product impact assessment standard in (#12), 74
    in program assessment principles, 253, 256
    students with disabilities and, 88
    teaching skills standards of, 264–266
    teaching to, 9
    technological literacy dimensions in, 28, 158
    tests and, 34
Stanford Education Assessment Laboratory (SEAL), 203, 209, 223–224
Starkweather, K., 17
statistical software, 117–118

Index

Steinberg, L. S., 211
"STEM" education (science, technology, engineering, and mathematics), 5
Stevens, A. L., 207
Stiggins, R. J., 4, 11, 33, 107, 109
Storm, G., 91
strategic knowledge, 204, 207–208
Sugrue, B., 205
summative assessments, 6, 129–132, 133, 257
Sylva, K., 188

T

Taggart, G. L., 242
talk-aloud protocols, 218
Task Group on Assessment and Testing (U.K.), 162
tasks, in assessments, 190–191, 211, 213, 222
Taylor, A. R., 160
Taylor, C. S., 212, 221
Taylor, J. S., 96
teacher performance, 3, 273–289
    assessment framework for, 285–287
    conducting assessments of, 264–266
    in-service assessment of, 284–285
    INTASC on, 276–278
    National Academies and, 29
    preservice assessment of, 282–284
    as reflective practitioner, 133–135
    teaching to standards and, 9
    teaching to tests and, 65
    technology education assessment of, 278–281
    university assessment of, 275–276
Technically Speaking: Why All Americans Need to Know More about Technology (National Academy of Engineering and National Research Council), 18, 26, 170
Technological literacy, 4. *See also* Standards for Technological Literacy (STL, International Technology Education Association)
    in academic disciplines, 29
    dimensions of, 26, 28
    National Academies and, 19–23, 28
    National Academy of Engineering and, 18
    "probe studies" of, 25
technology education, 1, 9–10, 278–281

Technology Education Research Unit (TERU), Goldsmiths College, University of London, 181, 195
Technology Enhanced Learning in Science center, 27
technology student associations, 266
Tech Prep programs, 235–236
Tech Tally (National Academy of Engineering and National Research Council), 2, 19, 22
Tedesco, M., 93
Tests. *See also* alternative assessment tools; classroom-based assessments; Educational Testing Service (ETS); large-scale assessments
    ACCUPLACER and ASSET, 143
    government, 25. *See also* National Assessment of Educational Progress (NAEP)
    high-stakes, 5, 7, 11
    performance on multiple, 140
    standardized, 11
    standard procedures for, 221
    students with disabilities and, 91–92
Tests in Print VII (Murphy, Spies, and Plake), 166
Thatcher, M., 182, 198–199
think-aloud protocols, 217
Think/Pair/Share, for self-assessment, 128
Third International Mathematics and Science Study (TIMSS), 159–160, 170, 205, 208
Three-Minute Pause, for self-assessment, 128
Threlfall, P. M., 190
timing of assessment, 8
Tindall, G., 93
Tombari, M. L., 73
Towles, E. A., 93–94
trade and industrial (T & I) education, 3, 233–249
    assessment instrument design in, 240–246
    interrelationship of purpose, objectives, and assessment in, 234–237
    program accountability in, 246–247
    skills assessed in, 237–240
    as technology education root, 33
true-false test items, 37, 39–41

Index

U

U. S. Congress, National Academies and, 17
U. S. Department of Education, 28, 141, 236
Understanding American Industry Test (Blomgren), 169
Understanding Design & Technology (Department for Education and Science, U. K.), 190
University of British Columbia, 157, 166–167
University of California-Berkeley, 77
University of Colorado at Denver, 203
University of London, 19, 159, 162, 181, 195
University of Maryland Eastern Shore, 251
University of Minnesota, 118
University of Ottawa, 157
University of Saskatchewan, 162
University of Virginia, 186
unpacking standards, 74, 76
utility of rubrics, 72

V

validity
    in assessing students with disabilities, 89–90
    of assessment aligned with purpose, 7–8
    content, 53–54, 89, 106–108
    in data analysis, 105–108
    face, 89, 106, 108
    of group-oriented assessments, 94
    instructional, 73
    of multiple-choice test items, 41
    predictive, 89
    of rubrics, 72–73
    of standardized tests, 167
value of assessment, 10–14
van Rensburg, S., 161
Views of Science-Technology-Society (VOSTS) questionnaire (University of Saskatchewan), 162
Virginia Technological University, 160
Vocational Competency Achievement Tracking System (VoCATS), North Carolina Department of Public Instruction (NCDPI), 79, 139–156
    analyzing data from, 149–153
    assessment expansion of, 141–143
    case study on, 143–145
    development of, 139–141
    future of, 153–154
    scope and sequence and, 145–149
Volk, K., 161

W

Walberg, H. J., 65, 130
Ward, A. W., 242
warehouses, data, 115–117
Waxman, H., 168
Webb, N. M., 94, 218
Web-based electronic portfolios, 27
Weil, M., 166
Weiner, B., 127–128
Welfare, R., 3, 139
Welty, K., 169
"what-if" possibilities, in design, 189
Wheeler, T., 162, 186
White, R., 205
"wicked" problems, 188
Wiggins, G. P., 67, 108, 192, 241
Wiliam, D., 5, 219, 223
Williams, R., 183
Wood, M., 242
Woodrow, J., 165
Work, S., 276
workplace attitudes, 237
Wosniak, A., 162, 186
Wright, A., 165
Wulf, W., 17
Wynne, B., 159, 168

Y

Yin, Y., 205
Yip, W. M., 161
Young, D., 215
Young, T., 165, 170
Young Children's Computer Inventory, 166

Z

Zedech, S., 167
Zemelman, S., 113
Zuga, K., 170